IMAGES
of Aviation

WILLOW RUN

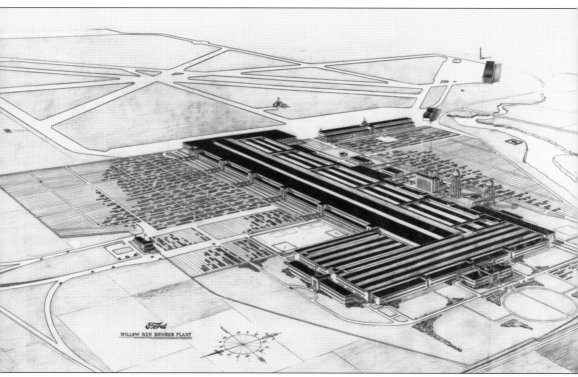

Here is an architectural sketch of Ford Motor Company's Willow Run Bomber Plant and associated airfield as it looked in the spring of 1945, when the last Ford-built B-24 was delivered. (Yankee Air Museum.)

ON THE COVER: On June 28, 1945, the last Ford-built B-24 bomber rolled out of the south-facing doors. Standing by the tractor as the airplane was being pulled out of the factory was Henry Ford II, who was to become president of Ford Motor Company. The airplane was surrounded by the last of the employees who pulled off one on the production miracles of World War II. Even before World War II started for the United States, Ford Motor Company had undertaken an almost impossible task: building four-engine B-24 bombers at the rate of one per hour on an automotive-style assembly line. As the Army stated after World War II, the question of whether the auto industry could be successfully converted to manufacture airplanes left no room for debate. (Yankee Air Museum.)

IMAGES
of Aviation

WILLOW RUN

Randy Hotton and Michael W.R. Davis

ARCADIA
PUBLISHING

Published by Arcadia Publishing
Charleston, South Carolina

Printed in the United States of America

Library of Congress Control Number: 2016931943

For all general information, please contact Arcadia Publishing:
Telephone 843-853-2070
Fax 843-853-0044
E-mail sales@arcadiapublishing.com
For customer service and orders:
Toll-Free 1-888-313-2665

Visit us on the Internet at www.arcadiapublishing.com

*This book is dedicated to Randy's father, Andy Hotton,
and to all the manufacturing legends of Detroit who
made this production miracle possible.*

CONTENTS

ACKNOWLEDGMENTS

This book would not have been possible without the help of my coauthor Michael W.R. Davis. Mike's experience in creating several books for Arcadia was an immeasurable resource for me in my quest to put this book together. The support of the Yankee Air Museum's executive director, Kevin Walsh, and access to its archives has been exceptional. Particularly Julie Osborne, the museum's curator, provided a treasure trove of historic pictures covering the history of the Willow Run Bomber Plant. I want to thank Ryan Jelson and Linda Skolarus at the Benson Ford Research Center at The Henry Ford in Dearborn, Michigan. I also want to thank Malgosia Myc, lead reference archivist at the Bentley Research Center at the University of Michigan, for her help in accessing the Albert Kahn blueprints for Willow Run. Again, thanks go to Thomas Allen and the staff of the University of Texas-Dallas Eugene McDermott Library for their help in gaining access to the Consolidated and Jimmy Doolittle collections. Feiga Weiss from the Holocaust Memorial Center in Farmington Hills, Michigan, was most helpful in obtaining needed photographs. A special thanks goes to Mark Bowen at the Detroit Public Library for his help. Particular thanks are extended to Charles Sorensen's grandson Charles Sorensen for his help and to Rose Monroe's daughter Vicky Croston for her insights into her mother's life at Willow Run.

All photographs used in this are courtesy of the Yankee Air Museum unless otherwise credited. The following is the key to the other photographic credits and sources:

(HF) From the collections of The Henry Ford, Benson Ford Research Center
(BL) From the Albert Kahn collection, Bentley Historical Library, University of Michigan
(LC) Courtesy of the Library of Congress
(FR) Courtesy of Franklin D. Roosevelt Library Collection
(CS) Courtesy of Charles Sorensen for the Charles Sorensen family
(USHMM) Courtesy of the United States Holocaust Memorial Museum. The views or opinions expressed in this book, and context in which the images are used, do not necessarily reflect the views or policy of, nor imply approval or endorsement by the US Holocaust Memorial Museum.
(NA) Courtesy of the National Automotive History Collection, Detroit Public Library
(TACOM) Courtesy of TACOM Archives, US Army, Warren

INTRODUCTION

I have been a member of the Yankee Air Museum located at Willow Run since 1985. This book is a snapshot of the story that the Yankee Air Museum will tell when it moves into the former Willow Run Bomber Plant.

My earliest memories of Willow Run go back to 1947 when I was four years old. My parents and I were driving out to Willow Run Airport for some reason, and my father pointed to this giant building that seemed to run the length of Ecorse Road. He said, "That is where I worked during the war. I built B-24 Bombers." Every time the family would drive by Willow Run, he would repeat this story. When veteran Ford Motor Company manufacturing expert Mead Bricker took over the Willow Run factory and instituted his 505 program, the outsourcing of small parts production became an improved mode of operation at Willow Run. In the 1930s, all Ford dealers had machine shops. These dealership machine shops became largely idle due to World War II wartime rationing of cars, gas, and tires. To utilize this untapped production potential, jobs were assigned to these dealerships, and my dad would set up and manage these projects.

Accounts of World War II are centered around the attack on Pearl Harbor, the great battles at Midway, the landings at Normandy, and the Battle of Bulge. There is no doubt that the heroics and dedication of our military men made these victories possible. But World War II was a war of machines. Those machines would come from American industry that was centered in Detroit. The manufacturing mastery developed in Detroit would save the world from Axis domination.

The United States committed to fight World War II with machines, and the machine that defined America's war effort was the four-engine bomber. In September 1940, the Battle of Britain had been won; the fighters flown by the RAF "Few" had defended England, but the four-engine bomber was already identified as the offensive weapon to carry the war to Germany. The auto industry would be become the backbone of four-engine bomber production. The airplanes would have names on them like Consolidated and Boeing. However, inside those bombers you would find parts with names on them like Ford, General Motors, Chrysler, Briggs, Nash, Hudson, Budd, Studebaker, Buick, and Delco-Remy. Ford Motor Company would build a factory to mass-produce four-engine bombers like cars. It would be built beside a sleepy little creek near Ypsilanti, Michigan, named Willow Run, and it would be the largest factory ever built up to that time. The bomber factory would later be called the Willow Run Bomber Plant and would become the manifestation of America's ability to outproduce anyone in the world.

As the European nations prepared for war, the United States elected to withdraw from world affairs. With the Depression in full swing, the Army Air Corps was a hollow force. The aircraft industry that supported the Army Air Corps was barely surviving on a thin dribble of orders. Soon, though, the White House would want airplanes by the 10,000s. In May 1940, President Roosevelt called for 50,000 military airplanes a year. He then "drafted" the president of General Motors, William Knudsen, to spearhead this production goal. Knudsen was appointed as the production expert on the National Defense Advisory Commission. He knew this lofty goal of 50,000 airplanes would not be reached without tapping the manufacturing mastery of Detroit's auto companies to produce aircraft parts.

But before that could happen, Knudsen knew he had to remove the obstacles to manufacturing that were centered in the entrenched bureaucratic policies found in Washington, DC. There was no time for the old contract processes; tax laws would need to be changed, and incentives given to make it attractive for companies to do business with the government. With removal of these obstacles, four-engine bomber production levels would grow over twentyfold in the next 18 months under Knudsen's leadership.

On September 9, 1940, the first US War Department aircraft production Schedule 8-A was published, under the guise of "Defense" work. Under this schedule, there would be 1,047 four-engine bombers built in the next 28 months, before the end the 1942, at a rate of about 47 per month. That September, Knudsen visited existing airplane plants, largely on the West Coast, finding US bomber production capability inadequate. He proposed to President Roosevelt that two government-owned factories be built to produce bombers. Congress then approved funds for the government aircraft factories. Gen. Henry H. "Hap" Arnold, chief of staff of the Army Air Corps, notified Knudsen that these plants should be built to produce four-engine bombers.

On October 4, 1940, Robert Patterson, assistant secretary of war, expressed concern about where the parts would come from to build airplanes. Knudsen took the lead and broke the stalemate; no one knew the productive potential of Detroit better than he did. And it was Knudsen rather than other auto executives or Army Air Corps officers who took the initiative of calling in America's automobile manufacturers. He addressed a meeting of automakers on October 15, 1940, at the New York Auto Show held at Madison Square Garden. There, he asked for their help in building the bombers. Patterson stated on October 18 that the four-engine bomber had to be the production focus, for only that bomber could exert power in the Far East because of the great distances it could go. Schedule 8-A was superseded on October 23, replaced by Schedule 8-B; this changed aircraft production goals with a shift to four-engine bombers. Patterson and Hap Arnold told Knudsen that the four-engine B-24 and twin-engine B-26 should be built at the two new factories recently approved by Congress. At that time, George Mead, head of the Aeronautical Section of the National Defense Advisory Commission, requested costs to build 100 Consolidated B-24 heavy bombers and 200 Martin B-26 medium bombers a month at the two new government factories. On October 25, Knudsen brought the powerful Automobile Manufacturers Association (AMA), aircraft manufacturers, and the Army Air Corps together at the New Center Building in Detroit. There, he laid the groundwork for building 12,000 additional bombers over the then current bomber production plan.

But in early November, it was determined that the two new factories approved in September would not to be able to build an additional 12,000 bombers by 1942. Mead told Knudsen on November 15, 1940, that Consolidated Aircraft and Martin Aircraft companies had agreed to sell the right to build airplanes to the auto companies. The auto companies would use their managerial and design talent to build subassemblies for the new bomber plants. President Roosevelt fully supported this program and approved the $1 billion project to build the bomber plants at Kansas City, Missouri, and Omaha, Nebraska. In a November 18 memo to Knudsen, more than a year before Pearl Harbor, Mead stated if further expansion was needed, the larger auto companies would have to be licensed to build complete airplanes. This laid the groundwork for building Willow Run and suggested all expansion in bomber production now would come from the auto companies. This included filling a tremendous need for aircraft engines. At this time, Hap Arnold issued a directive that all bomber factories should be designed big enough to build the larger B-29 or B-32 bombers still years away. This is why Willow Run has delivery doors that are 144 feet wide when the B-24 only had a 110-foot wingspan.

In December, Patterson stated there was too much work being scheduled at the two new factories and the shortage of workers would hinder production. He requested that the government construct two more bomber factories in order to build B-24s. The Kansas City plant was then assigned for the B-25 twin-engine medium bomber, and the B-26 twin-engine medium bomber went to Omaha as originally projected. In mid-December, Mead and Maj. James H. "Jimmy" Doolittle—famous for his aviation pioneering and later for his raid on Tokyo from the aircraft carrier *Hornet*—visited Ford, General Motors, and Chrysler. Each company was to provide parts for one of the airplanes—General Motor would supply B-25 parts; Chrysler, Hudson, and Goodyear would supply B-26 parts; and Ford would supply the B-24 parts. Ford agreed to visit Consolidated in San Diego after the first of the year. In late December, President Roosevelt told Knudsen he wanted the B-24 plants built in Fort Worth, Texas, and Tulsa, Oklahoma. The Fort Worth plant would be managed by Consolidated Aircraft and the Tulsa plant managed by Douglas Aircraft. Parts to build B-24s at these plants would come from Ford.

On January 6, 1941, Charles Sorensen, Ford's executive vice president of manufacturing, visited San Diego and made his famous promise, "Ford can build one bomber an hour." This is just what the Army was looking for, and Ford was asked to start the project as soon as possible. The Army told Patterson on February 11 that Ford would supply parts to build 50 B-24s a month for both Fort Worth and Tulsa. A year later, Willow Run would be asked to build 100 knockdowns plus 405 flyable bombers. The plan was to build a manufacturing facility at the Rouge Complex in Dearborn, Michigan, and a new subassembly plant in western Wayne County near Ypsilanti, Michigan, but on February 13, Mead recommend building both facilities near Ypsilanti. On February 21, a contract was signed to have Ford build parts and subassemblies for 50 B-24s a month each for Tulsa and Fort Worth. Ford started planning a new parts manufacturing and subassemblies factory near Ypsilanti, Michigan. The production schedule was again modified, and Schedule 8-C was issued on March 1; this schedule actually reduced production based upon realistic assessment of the manufacturing capabilities. On March 6, a contract was issued to build an Albert Kahn–designed, 858,000-square-foot Airplane Parts Manufacturing Building near Ypsilanti. Ford already owned large tracks of farmland at the plant site.

In April 1941, as rapidly as things were moving in the ramp-up of four-engine bomber production, President Roosevelt complained about the slowness of building bomber factories; he wanted more bombers. Production was again modified, and Schedule 8-D issued, increasing aircraft production. On April 14, 1941, due to pending changes in aircraft production, Ford's plans for the bomber parts manufacturing facility were put on hold by the government. In a War Department meeting on April 24, the Army disclosed it wanted 500 four-engine bombers a month by June 1943. On April 25, ground was broken on the Fort Worth, Tulsa, and Ypsilanti B-24 factories.

On May 6, President Roosevelt ordered the 500 bomber program to get going at once. The advance of the 500 bomber program made obsolete Schedule 8-D, issued just three weeks earlier, and was due to the complexity of building four-engine bombers. The production schedule showed that the industry was only capable of building 200 four-engine bombers a month. The only way to reach the higher production goals was to pursue the option that Mead and Knudsen had discussed the previous November: having the auto companies build complete airplanes. The increase in four-engine bomber production called for 65 subassemblies a month from Ford for both Tulsa and Fort Worth, plus 75 complete flyable airplanes to be built at the Ypsilanti facilities. Consolidated San Diego would build 90 B-24s per month, bringing the total to 310 B-24s a month by June 1943. On May 19, Ford was notified to plan on building complete airplanes. On June 2, 1941, Schedule 8-E was issued that called for 500 four-engine bombers. On June 30, working drawings were issued for the new Willow Run Bomber Plant with 2,829,100 square feet of production floor space.

However, the 500 bomber program would result in a shortage of engines. Knudsen again turned to the auto companies to build the R-1830s for the B-24s and the R-1820s for the B-17s. In June, Chevrolet and Buick were awarded contracts to build B-24 engines and Studebaker a contract to build B-17 engines. On July 19, 1941, Schedule 8-F was issued, and plans were made to raise four-engine bomber production to 700 a month by June 1943 by eliminating 4,577 airplanes from production to support the four-engine bomber program. Knudsen elected not to do anything immediately because he felt this would just confuse an already frenzied environment.

Focus on the four-engine bomber production continued to drive the aircraft production schedules. In October 1941, Schedule 8-G was issued that stretched out the delivery schedule of two-engine bombers with no change in production numbers to emphasize the four-engine bombers. By November, the Army wanted more four-engine bombers, and on December 1, 1941, a week before the Japanese attack on Pearl Harbor, Schedule 8-H was issued for 1,000 four-engine bombers per month by June 1944. Most of this growth was intended for the B-29 Superfortress bomber; however, a fifth B-24 factory was included to be operated by North American Aviation in Dallas, Texas. B-29 plants were planned for Cleveland, to be run by General Motors, and Atlanta, to be run by Bell Aircraft. Plans for the B-29 plant in Cleveland were cancelled, and instead the Omaha B-26 plant was converted to building B-29s in 1943. In June 1944, the combined aircraft and auto industries would produce 1,065 four-engine bombers, of which almost half came from Willow Run. On January 12, 1942, due to the attack on Pearl Harbor, the Willow Run production schedule agreed to only six months before was increased from 210 planes per month to 405 planes per month, 150 complete airplanes and 255 knockdown airplanes assemblies. On February 26, 1942, the production schedule for Willow Run was increased again to 505 airplanes per month; a year later this became the foundation of Mead Bricker's 505 program to fulfill Sorensen's bomber-an-hour goal. In terms of airframe weight, the B-24 would be the most-produced such airplane in aviation history—until 2012, when it was surpassed by the Boeing 737. This is even more amazing when you consider the B-24 was only in production for four years and the 737 has been produced for over 50 years.

The aircraft industry's limited resources could not have achieved the remarkable production record of the years 1943–1945 without Detroit. The United States would mass-produce over 34,000 four-engine bombers. The need for heavy bombers was foreseen in the fall of 1940, and that number was expanded many times. The role of William Knudsen is often downplayed in this miracle of aircraft production, but it was the period before Pearl Harbor when Knudsen, almost all on his own initiative, brought the auto industry's manufacturing mastery into four-engine bomber production. This laid the basis for the distinctive achievement. The additional resources of managerial and engineering talent—and, to a lesser extent, of machinery and facilities—came chiefly from the giant automotive industry, which to most Americans was the embodiment of the principle of mass production. In November 1941, the factories were under construction, but parts were already coming off the assembly lines. The bombers that would rule Axis skies in 1944 were already in production. Willow Run was the manifestation of four-engine bomber production representing the masters of production found in Detroit.

—Randy Hotton

One

HOW DETROIT
SAVED THE WORLD

Many might say that titling a chapter "How Detroit Saved the World" during World War II is not an accurate portrayal of World War II. Of course Detroit did not do it alone, but the intellectual and managerial skills found in American manufacturing were a result of the auto industry's perfection of mass production. There would be no invasion at Normandy without the machines that rolled out of American factories. The Russians, as admitted by Joseph Stalin, would not have survived the 1942 German onslaught without the outpouring of American machinery under the Lend-Lease Program. Throughout the world, up until the 20th century in Detroit, "craft production" was the normal way. Craft production is distinguished by producing in small batches or one at a time, high craftsmanship quality, skilled labor with intensive handwork, and custom, individual fitting of parts with general purpose tools, such as hammers and files. This resulted in low volume and high cost per unit. On the other hand, "mass production," as perfected in Detroit, was characterized by little handwork, interchangeable parts, specialized tools, automation, unskilled or semiskilled labor, and moderate quality. This allowed high volume and low cost. The perfection of mass-production capacity in the United States was centered in Detroit, Michigan. In Detroit, Henry Ford and the Ford Motor Company had pioneered and refined mass production with standardized parts and the moving assembly line. In Detroit, Alfred Sloan of General Motors had broken new ground with management and marketing ingenuity of rapid model changes found in the annual model changeovers. In addition, the auto industry had developed an extensive network of large and small suppliers and a huge workforce of experienced production workers, especially in critical tool and die skills. These innovations gave the United States the ability to outproduce the rest of the world combined and convert from civilian production to military production far faster than any of its partners or enemies. Without Detroit's perfection of mass production, there would have been no production miracle and the world could be a much different place. The four-engine bomber emerged as the Allied offensive weapon of World War II, especially for Britain and the United States. In November 1941, as war raged in Europe and before Pearl Harbor, the four-engine bomber program had been launched, and the auto industry had become the primary supplier of bomber engines and airframe subassemblies.

The B-24 showed great promise in 1941, and the Army wanted Liberators by the thousands. The B-24 bomber's purpose was to rain destruction upon the enemy. American industry produced over 19,526 B-24s. By 1945, the B-24 had practically passed unmentioned into history. Shown here is a crew with its B-24. 2nd Lt. Edward Ratkovich, back row, second from the left, was a pilot of this B-24. Born in Detroit, he graduated from Cass Technical High School. He earned his pilot wings and was assigned to the Fifteenth Air Force in Italy. In 1945, his crew flew a dozen combat missions over German-occupied Europe in a B-24 Liberator built at Willow Run. Tens of thousands of crews like this one would become the pointy end of the spear that made the B-24 a liberator of occupied countries. This book is about the B-24, but without the crews who operated all over the world—from skies over Europe to the North Atlantic antisubmarine patrols to long-distance island raids in the Pacific—the B-24 would have been just an expensive collection of parts. These crews made a difference and must never be forgotten. (Michael W.R. Davis.)

A photograph of Henry Ford shows him with his first car, the 1896 Quadricycle, and his famous Model T. Over three years, in his spare time, he built the prototype Quadricycle in a workshop behind his home in downtown Detroit. Ford was a gifted mechanic who worked for Edison Illuminating Co. (today's DTE) as chief engineer. As an interesting side note, due to the invention of the telephone, he stood what is known today as a "phone watch" in his workshop in case there were problems at the steam-powered Edison generating facility a few blocks away. (HF.)

Shown here in 1906 is the Ford Piquette plant, where such early models as the R were hand assembled using the labor-intensive method of bringing parts to the car. It was a slow, laborious, low-volume, high-cost process, making custom-finished parts that were rarely interchangeable. Ford was building about 25 cars a week at this time. (HF.)

Ford's desire was to build a "miracle" car that everybody could own and drive. Being that most cars to that time had been custom-built or built one at a time in limited quantities, Ford had the vision to implement and successfully demonstrate how he could revolutionize the assembly process by using identical interchangeable parts. Ford's perfection of standardized parts, unskilled labor, and the moving assembly line revolutionized manufacturing. No one else in the world could produce cars at a lower cost. (HF.)

By 1913, the Ford Highland Park plant, shown here, would be producing a thousand Model Ts a day. By using moving belts, workers could remain at one location and do one task well. In an attempt to literally put a Ford in every garage, Ford was able to bring the cost of a Model T down to $290 by 1915, a year in which he produced nearly half the world's automobiles. (HF.)

Shown here on the right is Alfred P. Sloan Jr., president of General Motors (GM) from 1923 to 1935. By the mid-1920s, initial demand for the basic car had been filled. Now, buyers wanted more features and more prestigious automobiles. GM introduced stair-step marketing and an annual model change with a variety of models. This required tremendous flexibility and rapid, though costly, changeover to introduce new cars. This forced Ford's standardized parts and moving assembly line system to the next level and gave great flexibility to the Army with its rapidly changing designs for airplanes and armored vehicles. (LC.)

This photograph of the Auto Show in Louisville, Kentucky, in 1931 shows off the General Motors car lines. GM's stair-step marketing had five different products, starting with the low-priced Chevrolet, up to the slightly higher-priced Pontiac, on to Oldsmobile, then to Buick, and finally to the top-line Cadillac. With value-added features to justify carefully layered ascending prices, the concept was designed to keep the motorist in the GM family. (LC.)

Shown here are the beaches of Normandy, France, where the Allies landed on June 6, 1944. It is littered with machines that poured out of the "Arsenal of Democracy" perfected in Detroit. Gen. Dwight Eisenhower said when looking down at the beaches that "any military commander would be devastated by the loss of the equipment seen on those beaches, but I knew I had five times as much material in England waiting to come over." As Stalin said, "This is a war of machines, United States is a country of machines, the country with the most machines will win the war." (LC.)

The Reo 2.5-ton truck shown here is similar to the Studebaker six-by-six truck. Two hundred thousand of these trucks were shipped to Russia under the Lead-Lease Program. The United States alone provided the Russians with 501,660 tactical wheeled and tracked vehicles, including 77,972 jeeps; 151,053 one-and-a-half-ton trucks; and 200,622 two-and-a-half-ton trucks. The aid was vital, not only because of the sheer quantity, but because of the quality, drivability, and durability. (TACOM.)

The B-24 shown here was the most-produced bomber in World War II. In the fall of 1941, the four-engine bomber emerged as the Allied offensive weapon that would take the war to the enemy's home cities. US companies and their suppliers would be the source of the engines, subassemblies, and parts that would allow the United States to build over 34,000 of these airplanes.

Ground was broken along Willow Run creek near Ypsilanti, Michigan, in April 1941. Six months later, in October 1941, two months before Pearl Harbor, the plant was already producing parts. Auto company manufacturing methods would allow the United States to build a massive fleet of B-24 bombers that would rule the skies throughout the world.

Andy Hotton, coauthor Randy Hotton's father, was a graduate of the Ford Trade School. He was a toolmaker at Willow Run. He was hired in the winter of 1942 and worked there until three days before the last B-24 was delivered. He is shown with his 1939 Ford hot rod that had a modified bored-and-stoked V-8 engine. In 1940, eighty-five percent of the cars in the world were registered in the United States. Americans tended to be mechanically oriented. Tinkering with automobiles and their engines, young American men seemed to be particularly adept at mechanical devices and solutions. This produced a generation of people literate in necessary mechanical skills. These "high school hot-rodders" knew how to build things and fix machines. This was one of America's secret weapons during World War II. (Author's collection.)

Two

ISOLATION FROM THE WORLD

The United States failed in its World War I mobilization efforts. President Wilson appointed a panel of progressives with virtually no industrial experience to mobilize the United States. The commission faltered and was superseded by a War Industries Board (WIB). When WIB's mobilization effort also collapsed, President Wilson appointed Bernard M. Baruch, a well-known Wall Street financier, to head the WIB. In late 1918, Baruch had successfully managed the American economic mobilization, but by that time, World War I was over. Franklin D. Roosevelt, assistant secretary of the Navy then, keenly observed that these initial efforts were unfocused and misdirected and that Baruch had brought order to the mobilization. This encounter would lead to a much different type of mobilization when World War II came. Following World War I, America was anxious to demobilize its military forces rapidly, as it had done after every war in the past. America's once formidable arms production capability fell into disuse and eventual liquidation. Congress passed laws encouraging weapons producers to get rid of their arms manufacturing capability.

Despite the world suffering from an economic depression, the drums of war began to roll. In 1933, Adolf Hitler came to power in Germany. He started building a national "Wartime Economy" in pursuit of an industrial age total war. In 1937, as the European nations played the game of appeasement with Hitler, Congress passed isolationist laws to keep the United States disengaged from world affairs. President Roosevelt asked Congress to build up the Army Air Corps as a way to prepare the United States for a possible conflict in Europe. He felt that airpower would be the way to send a message to the Germans. Initially, Congress rejected Roosevelt's request to airpower. In November 1938, the appeasement of Germany resulted in the first overt outrage against the Jews in Germany. The next day, Roosevelt asked Congress for more military airplanes. He knew that building a new Army fort in Wyoming would not impress potential enemies, but a powerful air force might prevent the next war. The Germans invaded Poland in September 1939. The airplane-manufacturing program, which held such promise in the spring of 1939, now seemed woefully inadequate in the light of the Luftwaffe's success during the Polish invasion.

Under the scholarly leadership of President Wilson, the United States failed in its mobilization efforts during World War I after declaring war in April 1917, when Franklin Roosevelt became assistant secretary of the Navy. Neither Roosevelt nor the Army had experience with big business. The independent, decentralized bureaus at the heart of the War and Navy Departments' supply system, a leftover from Indian-fighting days, often seemed more adept at defending their individual prerogatives than supplying a large Army. (LC.)

The War Industries Board (WIB) shown below was supposed to encourage companies to use mass-production techniques to increase efficiency and urge them to eliminate waste by standardizing products. The board in Washington set production quotas and allocated raw materials. When this mobilization effort also collapsed in the winter of 1917–1918, Wilson appointed Bernard M. Baruch, a well-known Wall Street financier, as head of the WIB. (LC.)

In 1916, Baruch, shown on the lower right, had left Wall Street to advise Pres. Woodrow Wilson on national defense. After appointment to the WIB, Baruch successfully managed America's economic mobilization during World War I. But by the time he had organized US industry, World War I was over. The actual production record prior to the armistice was paltry at best. US troops in Europe fought the war with equipment supplied by France and Britain. (LC.)

Seen here at a gathering of Navy Department officials in 1916, Franklin D. Roosevelt (second from left on the first row) served as assistant secretary of the Navy during World War I. He observed firsthand the failure of the Wilson administration to mobilize the United States. He also observed the steady hand that Bernard Baruch brought to the mobilization effort. Roosevelt and Baruch maintained a relationship after World War I, and Baruch became one of his closest advisors. (LC.)

Following World War I, many Americans clearly sought a return to quieter times and more traditional values. Military contracts were canceled on almost no notice, and the empty factory shown here was a result of these cancellations. With empty factories, many companies lost money on their investments for the government and would hesitate to enter into military contracts in the future. Politicians carried their constituents' sentiments to the House floor. When these sentiments were applied to the US Army, it meant a small and inexpensive force composed of volunteers, far from the sight and mind of the general public. By 1921, Congressional appropriations had cut the regular Army to 125,000 due to public pressure and economics. (LC.)

When the Allies were fighting World War II in North Africa in the spring of 1943, Secretary of War Henry Stimson, shown here, testified before Congress and reflected on the desperate state of the US military preparedness prior to World War II: "In the spring of 1940, we did not have enough powder in the whole United States to last men we now have fighting overseas for anything like a day's fighting, and what is worse we did not have the powder plants or facilities to make it; they had all been destroyed after the last war." (LC.)

In 1928, Joseph Stalin (right) came to power in the Soviet Union. He wanted to build the strongest military in the world. Stalin set production goals of 100,000 airplanes and 100,000 tanks. In 1931, Japan also started to arm and invaded Manchuria to ensure supplies of raw materials for its war industries. The arms race had begun. (LC.)

Long before Pearl Harbor, Washington became concerned that the mobilization plan for the United States must be updated should America again be forced to fight in another world war. In 1931, under the guidance of Gen. Douglas MacArthur, a major on the Army staff named Dwight Eisenhower was given the task of reviewing the country's mobilization plans. He spent a year reviewing, studying, and interacting with America's then weak military-industrial complex. Eisenhower recommended that "Education Orders" be given to manufactures to test the mobilization plan, but no such orders were placed. (USHMM.)

Shown here is Adolf Hitler, at left, with Field Marshal Hermann Göring, second from left. In 1933, when Hitler came to power in Germany, he scolded his ministers for not spending money fast enough on arms. He started building up the German army in pursuit of the world's mightiest military. In 1937, Göring made plans for a Luftwaffe of 7,900 airplanes. Airpower would define the next war. (USHMM.)

In 1936, the Senate passed the Neutrality Acts sponsored by Sen. Gerald Nye, shown at left. The world was well aware of the production potential of the United States. But America was an isolationist nation in the 1930s and did nothing. This was to ensure that the arms manufacturers, known as "Merchants of Death," would not be allowed to make and sell arms overseas. In theory, this would prevent America being drawn into another European war. (LC.)

The appeasement of Germany resulted in attacks against the Jews in that nation. Known as Kristallnacht, or "Night of Broken Glass," this was a series of coordinated attacks against Jews throughout Nazi Germany. The attacks left the streets covered with broken glass from the windows of synagogues and Jewish-owned homes and stores, such as the leather goods store shown here. (USHMM.)

On November 12, 1938, the day after Kristallnacht, Roosevelt asked Congress for funding to build 10,000 airplanes. Congress debated for seven months before approving just 5,500 airplanes for the Army and Navy over the next five years. But Congress elected not to fund the Boeing B-17 bomber, the world's most formidable bomber, because it was considered an offensive weapon. (LC.)

The Germans invaded Poland in September 1939, typified by this photograph of a German solider near Warsaw. America's 5,500 airplane program, which held such promise in the spring of 1939, now seemed woefully inadequate in the light of the Luftwaffe's success during the Polish invasion. Airpower that played a major role in the defeat of the Polish army now became the key to victory in World War II. (USHMM.)

Three

50,000 AIRPLANES

On May 10, 1940, the Germans invaded the Low Countries—Belgium, Luxembourg, and the Netherlands—and France. This was the blitzkrieg, combining armor, mechanized infantry, and airpower. Airpower was now decisive. The blitzkrieg brought about a meeting between Roosevelt and Gen. George C. Marshall, chief of staff of the US Army. Marshall alerted the president that the United States needed to get serious about rearming efforts. A bold step was needed to expedite the buildup of US airpower for the Army Air Corps and Navy. On May 16, 1940, Roosevelt addressed Congress and made his famous speech that called for 50,000 airplanes per year. When Hermann Göring, head of the Luftwaffe, heard this figure, he laughed and said, "This is pure propaganda, because no one could build 50,000 airplanes a year." The United States would not only produce 50,000 airplanes per year, but also exceed that number in 1944 by producing nearly 100,000 airplanes. On May 23, 1940, Roosevelt sensed that his options had run out. He called the one person he trusted to mobilize America's productive potential, and that was Bernard Baruch. But Baruch turned him down; at age 69, he felt he was too old for the complicated job. "Someone else," he told Roosevelt, "would have to take charge." The president wanted to know whom he should call and asked Baruch, "Who are the top three industrial production men in the United States right now?" "First, second and third is Bill Knudsen," Baruch replied. Knudsen, the president of General Motors, was about to be asked to marshal Detroit's mass production to save the world. Knudsen knew the natural productivity of American capitalism firsthand. No one knew the productive potential of Detroit better than Knudsen did. He was on a first-name basis with all the titans of the US auto industry. Roosevelt called Knudsen and said, "I want to see you in Washington. I want you to work on some production matters. When can you come down?" Knudsen answered, "This country has been very good to me, and I want to pay it back. I will come down in two days." In their first meeting, he told Roosevelt, "I'm no soldier, but I know if we get into war, the winning of it will be purely a question of material production. Once I know how much you want and when do you want it, then we can start." At that time, no one could answer his question.

The Germans invaded France, the Netherlands, Belgium, and Luxembourg on May 10, 1940. This was a blitzkrieg combining tanks, mechanized infantry, and airpower, especially Stuka dive-bombers. In 10 days, the German army defeated the combined armies of France, England, Belgium, and Holland. (USHMM.)

In May 1940, the United States had the 18th-largest air force in the world—right behind Romania. The US Army Air Corps was made up of 326 "modern combat airplanes" that were obsolete by world standards, such as the P-39 shown here. The P-39 would be driven from the Pacific skies by Japanese Zeros following Pearl Harbor. There was no longer time for the peacetime bureaucratic process of airplane procurement. A bold step was needed to build up US airpower.

Shown here are General Marshall (left) and Secretary of War Henry Stimson reviewing a copy of Emergency Supplemental Appropriations. As the German blitzkrieg progressed in Europe, Chief of Staff of the Army George Marshall went to President Roosevelt and said, "We have to get together a group of industrialists to draw up plans for defense preparation and production. There is not a day to spare." Peacetime bureaucratic processes of airplane procurement no longer would allow the country to defend itself. (LC.)

In May 1940, after Germany's blitzkrieg, President Roosevelt addressed Congress, calling for authorization for 50,000 military airplanes per year. But he would have to turn to Detroit to solve the aircraft production bottleneck. When Hermann Göring, head of the German Luftwaffe, heard of this number he said that it was "pure propaganda . . . no one could produce 50,000 airplanes in a year." The United States would not only build 50,000 airplanes per year, but also exceed that number in 1944 by producing nearly 100,000 airplanes. (FR.)

Roosevelt realized that Washington alone could not manage a huge increase in "defense" production. So he called on World War I's Bernard Baruch, shown on the right, for advice. Baruch recommended the world's leading businessman, Bill Knudsen. Roosevelt was afraid of an adverse reaction from organized labor due to clashes that had happened a few years before between Knudsen and the unions during organization drives in Flint, Michigan. Baruch persisted, and now running out of time, Roosevelt finally agreed. (LC.)

Bernard Baruch had recommended Bill Knudsen, the president of General Motors, for the job of mobilizing the United States. Knudsen, shown here testifying before Congress in 1938, was a Danish immigrant who came to this country in 1900. He was a machinist who worked his way up from the shop floor into management at both Ford and General Motors. He was widely esteemed in the auto industry. (LC.)

Bill Knudsen is shown here in Washington, DC, on May 29, 1940. He had come to meet Pres. Franklin D. Roosevelt, who had appointed him as the production expert on the National Defense Advisory Commission. Knudsen knew that producing aircraft and airplane engines was the biggest manufacturing challenge facing the nation and that production goals would not be reached without tapping auto company mass-production manufacturing mastery. It took 18 months to get the machine tools in place to build the weapons to defend the United States. (LC.)

The Axis had a four-year head start in military production, possibly making US mobilization irrelevant. However, because of the production miracle perfected in Detroit, the US buildup for World War II took just over a year. This rapid mobilization, combined with America's strategic advantage in industrial infrastructure, workforce, and raw materials, eventually placed the Axis nations in a competition they could not win. America also had the manpower reserves to organize Army divisions, Navy fleets, USMC raiders, and a growing Army Air Corps. (LC.)

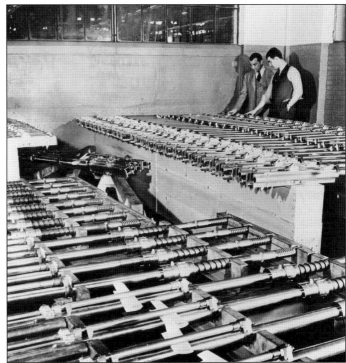

Shown here in the summer of 1940 is a meeting of the National Defense Advisory Commission (NDAC). The NDAC members were meeting to consider a Portative Program for Automotive War Production. The NDAC began as an emergency measure during World War I to promote industry and government coordination on war-related projects. Assistant Secretary of the Navy Franklin D. Roosevelt wrote then that he "heartily endorsed the principle" on which the legislation was based. In May 1940, after the German invasion of the Low Countries and France, Roosevelt resurrected this largely forgotten organization to bring business leaders into the nation's mobilization efforts and avoid going to Congress for permission to form a commission. Bill Knudsen, president of General Motors and a specialist on production, was appointed as head of the NDAC. The NDAC had no real power because it was an advisory commission. (LC.)

Four

KNUDSEN AND THE AUTO COMPANIES

Bill Knudsen had been appointed head of production for the National Defense Advisory Commission (NDAC) to prepare the United States for World War II. Knudsen knew the country was desperately short of machine tools to build the material needed to fight a war. There was no incentive for businesses to enter into contracts with the US government. Congress would not fund the expansion needed for mobilization. Industry was hesitant to fund these projects because of doubts about being able to ensure a return on investment and because of the experience following World War I when contracts were cancelled with no warning. Knudsen knew the potential of America's industry, but he also had to persuade Roosevelt to remove the barriers holding back production. He identified three obstacles preventing businesses from full voluntary participation in mobilization. The first was the laborious multilayered contract process that would have to be suspended and the process sped up by issuing letters of intent. The second was to scrap many of the New Deal's antibusiness regulations and tax policies. The third was to take the control of the war mobilization process away from Washington bureaucrats. Building things would have to be attractive to businesses in order for them to fully participate. Roosevelt realized that in a capitalistic economy, the ability to make profits was the motivator for businesses to produce, so he followed Knudsen's advice. In the Battle of Britain, bombers emerged as the offensive weapon that gave airpower the ability to win wars. Bombers became the way the Allies would fight World War II. The urgency to kick-start the building of bombers was now staggering. On October 15, 1940, Knudsen went to the New York Auto Show at Madison Square Gardens and asked the auto manufacturers to meet him in Detroit in two weeks. The auto men gathered again in Detroit at the New Center Building. Once inside the building, they met a panel Knudsen had assembled, composed of Air Corps officers and aviation executives. What Knudsen's Detroit meeting set in motion over the next five years would not only save America, but save the world. The bomber plan was gaining momentum. Further study in December 1940 revealed that it would be impossible to reach this level of bomber production in the two new plants then planned for Kansas City and Omaha. As the bomber program took shape during the late fall, it was decided to construct two additional bomber assembly plants for operation by aircraft companies with subassembly supplied by the auto companies.

Shown here is a bomber being built before the auto industry became involved. The first bomber production, Schedule 8-A, was issued on September 9, 1940. It called for 1,047 bombers to be built by the end of 1942. With the contributions of the auto industry, this schedule was about to start changing. (LC.)

In September 1940, Knudsen visited aircraft factories around the country and found their production capacity totally inadequate. After these visits, he proposed to President Roosevelt that the government build two factories to be owned by the government and operated by private business, especially aircraft companies. Funds for these plants were approved by Congress before the end of September. Shown here is one of the two factories built by the government, the Kansas City B-25 factory. Here, North American Aviation assembled B-25s with parts made by General Motors. (LC.)

By October 1940, many of the bureaucratic obstacles impeding the building of bombers had been removed. Under Knudsen's initial plan, the auto companies would become suppliers to the airframe manufacturers. Shown here with Bill Knudsen at the New York Auto Show are the leaders of the auto industry. As the auto show dinner ended, Knudsen stood up and said in his broken Danish accent, "We must build big bombers. We need more bombers than we can hope to get. You have got to help. I want to meet with you soon in Detroit." (NA.)

Shown here is Detroit's New Center Building, then a block from GM headquarters. Here, the automobile manufacturers assembled again on October 29. All the leaders of the auto industry were there. It was a cross section of Detroit's mass-production base: auto companies, body companies, industry suppliers, and tool and die companies. The auto industry found Knudsen's proposal immediately appealing, because it came from someone who truly understood the capabilities and limitations of the auto-manufacturing sector.

At the government's request, the industry subsequently established in Detroit an Automotive Committee for Air Defense and began surveys of available plants in order to determine what it could contribute to the program. In November 1940, Maj. James H. Doolittle, shown here second from the right, with Ford's Charles Sorensen, second from the left, was stationed in Detroit by the Air Corps as its liaison officer. He assisted the industry in its planning and played a significant role in Ford's decision to build the Willow Run Bomber Plant.

Shown here is the interior of the Graham-Paige auto plant on Detroit's far west side. Doolittle had laid out blueprints, plus a series of engine and equipment parts, for the auto executives to examine. "Here are the parts we need," he said in closing. "Pick out the ones you can build; I will get you a contract." Detroit was about to be turned loose to mass-produce bombers and overcome potential production bottlenecks. (NA.)

Douglas Aircraft and Ford Motor Company were selected to join with Consolidated Aircraft of San Diego, California, to produce 100 B-24s per month above the production planned for Consolidated's San Diego plant. Under this plan, Ford was to make parts for 100 knockdown sets of B-24s per month. Shown here are the knockdown kits envisioned at that time. Rapidly increasing demand for four-engine bombers would lead to Ford building complete airplanes, as well as knockdown kits.

The Willow Run Bomber Plant, shown here, has doors that are 144 feet wide, while the B-24 only had a wingspan of 110 feet. In November 1940, Gen. Hap Arnold, chief of the Army Air Corps, issued a directive that all bomber plants built in the future had to be large enough to accommodate B-29 and B-32 bombers. If England had fallen to the Germans, Willow Run might have produced the B-32 bomber. The Omaha, Nebraska, plant would later have to be converted from building B-26s to B-29s.

CIVILIAN PRODUCTION ADMINISTRATION
BUREAU OF DEMOBILIZATION

AIRCRAFT PRODUCTION POLICIES
UNDER THE NATIONAL DEFENSE ADVISORY COMMISSION
AND OFFICE OF PRODUCTION MANAGEMENT

MAY 1940 TO DECEMBER 1941

HISTORICAL REPORTS ON WAR ADMINISTRATION: WAR PRODUCTION BOARD

Special Study No. 21

23-991 Cover

Over the next 19 months, the NDAC (National Defense Advisory Commission) and later OPM (Office of Production Management) guided the fulfillment of Roosevelt's 50,000 airplanes per year production goal. Late in May 1940, the president appointed an Advisory Commission to the NDAC. While it had no real power, it was given the responsibility of arming the nation for defense. The broad area of industrial production was assigned to Knudsen, president of General Motors, who was widely recognized as a production expert. Immediate direction of aircraft production fell to George Mead, former vice president of United Aircraft Corporation, who in early June 1940 became the head of the aeronautical section of NDAC. In the 19 months under the guidance of NDAC/OPM, four-engine bomber production goals grew from 47 four-engine bombers per month to more than 1,000 four-engine bombers per month. (LC.)

Five

SORENSEN AND THE BOMBER PLANT

The B-24 was a problematic airplane. Ruben Fleet, head of Consolidated Aircraft, had convinced Hap Arnold, chief of the Army Air Corps, that he could build a bomber superior to the B-17. The new bomber would be called the B-24. It showed tremendous promise: it was faster, carried more, and flew farther than the B-17. The Army wanted lots of B-24s, and the Army knew Consolidated needed help to increase production of the B-24. In December 1940, Dr. George J. Mead and Maj. Jimmy Doolittle visited Henry Ford. One of the reasons for approaching Ford was that Ford had the largest engineering and design facilities in the world at its Rouge Complex and could be the answer to the B-24 production bottleneck. Mead and Doolittle got no objection from Henry Ford and found enthusiastic support from both Edsel Ford and Charles Sorensen, head of Ford production. In January 1941, Dr. Mead, Edsel Ford, Sorensen, and other key Ford production people visited Consolidated's San Diego plant to determine the feasibility of producing parts for the B-24. While there, Sorensen observed the major aspects of Consolidated's production with disdain. Sorensen told Fleet he neither liked what he saw nor what he heard, and said, "All this is pretty discouraging." Fleet looked at Sorensen and said, "So how would you do it?" Sorensen answered, "I don't know but I'll have an answer for you in the morning." Sorensen was staying at the Hotel del Coronado, across the bay from San Diego. With his notes from the day and his 35 years of manufacturing experience, he broke the design of the B-24 down into subassemblies. He figured out average job performance, time flow, and overhead costs. By 4:00 a.m., when he had the proper sequence down and production time allotted, he then sketched out the floor plan of a factory that would produce one B-24 bomber per hour. In the morning, Sorensen went to Fleet's office with his papers under his arm. He told Fleet that Ford could build a B-24 bomber in just an hour on an automotive-style assembly line. Fleet was dazed by the scale of Sorensen's proposition. He said, "You are crazy; no one can do that." This was a challenge from Fleet that Sorensen could not let stand. Sorensen had just taken on the biggest project of his life when he committed Ford to building a four-engine B-24 bomber at a factory that would be named the Willow Run Bomber Plant.

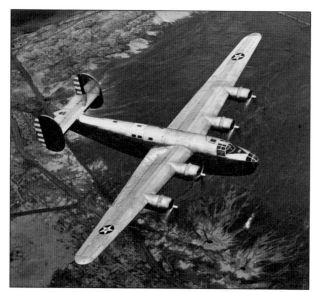

Shown here is the B-24A as designed by Consolidated, with its high-lift Davis wing, Fowler flaps, and tricycle landing gear. A revolutionary airplane design, the experimental XB-24 first flew in December 1939. It would fly farther, faster, and carry a bigger bomb load than the B-17. The Army Air Corps desperately wanted this airplane, but Consolidated's production was lagging, with only seven B-24s built in all of 1940. The B-24 was still in the prototype stage.

Ford's Rouge Complex was the largest vertically integrated production facility in the world. In mid-December 1940, Dr. George Mead, president of the US National Advisory Committee for Aeronautics (NACA), and Maj. Jimmy Doolittle went to visit the elderly Henry Ford. They wanted to determine his willingness to manufacture airplanes for the US defense program. Earlier in 1940, he had refused to build aircraft engines for the British. One reason for approaching Ford was because of his company's enormous purchasing, engineering, and production resources, possibly the answer to the perceived B-24 production bottleneck.

Shown here are Ford Motor Company president Edsel Ford (left) and Charles Sorensen (right), Ford's manufacturing chief, breaking ground for the bomber plant at Willow Run. They had gone to visit the B-24 headquarters in San Diego in January 1941. They wanted to see how B-24s were being built to determine the feasibility of Ford producing parts for B-24s. While there, Sorensen observed major aspects of Consolidated's production with disdain. No two airplanes were identical in their parts, there were no interchangeable parts, and Consolidated, according to Sorensen, did not even have blueprints to share with him. (HF.)

Pictured on the far shore in this photograph, the Consolidated airplane factory was built on marshlands at the inlet to San Diego Bay. So unstable was the ground that it would rise and fall with the movement of the tides in and out of the bay. This made it nearly impossible to make identical airplanes with interchangeable parts the way Ford planned to build components for B-24s. (LC.)

This B-24's final assembly at San Diego was made outdoors on steel fixtures. The heat and temperature change so distorted these fixtures that it was impossible to turn out two planes exactly alike. Consolidated worked in tolerances of 1/32 of an inch, while Ford worked in 1/1,000 of an inch. Sorensen was worried that any wings or fuselages built by Ford would never be compatible with fuselages constructed by Consolidated. (LC.)

Sorensen was staying at the famous Hotel del Coronado, shown here, across San Diego Bay from Consolidated. After dinner on the day of his tour, he went back to his room. He took his notes from the day and, with his 35 years of manufacturing experience broke the B-24 down into subassemblies. He figured out a variety of automotive-style production flows. (LC.)

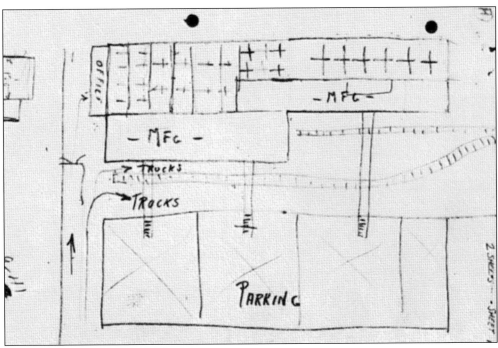

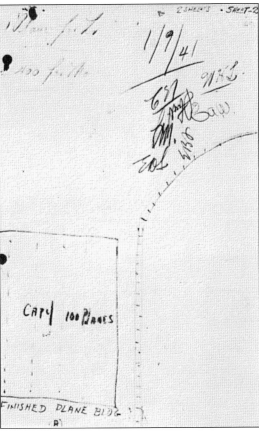

Shown here are the drawings that Sorensen sketched out in pencil on hotel stationery at 4:00 in the morning. On the right side of the photograph at right, Edsel Ford's initials can be seen by the date 1/9/41. The image above shows the floor plan of a factory that could produce a complete flyable B-24 at the rate of one per hour. In Sorensen's imagination, there would be no mere parts factory—it would become the vortex of President Roosevelt's Arsenal of Democracy. At breakfast, he showed Edsel Ford the sketches he had made. Ford was dazzled by the concept and urged him to go see the Consolidated people at once. He was elated by the certainty that the Germans had neither the facilities nor the concepts to mass-produce planes in this way. The sketches were shown to the Army, and officials were so impressed that they told Ford to start the project immediately. (Both, CS.)

Shown in the lower left of this picture is a sleepy little creek 26 miles west of Detroit, near Ypsilanti, Michigan. It was locally known as Willow Run. The floor plan sketched out on hotel stationery would shortly become an aircraft plant to be known as the Willow Run Bomber Plant.

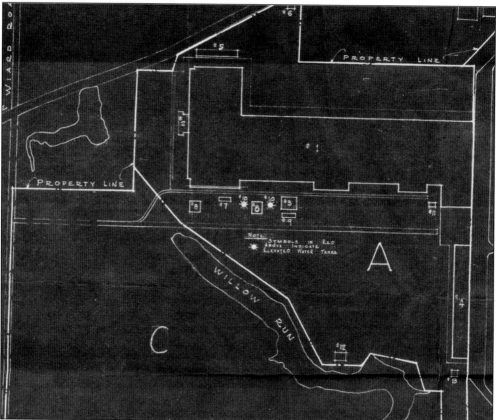

This blueprint shows the factory and adjoining four square miles of open land to be converted to what became the world's largest airport at that time. But building one B-24 every hour would be a far greater challenge than anyone could have imagined. When Ford was asked to undertake production of just B-24 parts, it was still a newly designed plane, not thoroughly tested or production engineered.

Six

BUILDING A
BOMBER PLANT

With the Army's approval to proceed, Charles Sorensen showed the elder Henry Ford his bomber plant plans. Ford said, "Make the complete airplane." This was Sorensen's authorization to proceed with building the world's largest manufacturing plant. A meeting was set in Washington, DC, on February 21, 1941, to issue the letter of intent. Sorensen had two things he had to do before that meeting: One, he needed to find three to four square miles of sparsely populated land for a factory and an airfield. He found this in western Wayne County. Two, he needed someone to design this new bomber assembly plant. For this job, Sorensen contracted with Kahn Associates, the architectural firm he had dealt with since 1910. As a result of the February 21 meeting, Ford received a $200,000,000 contract to build 1,200 fuselage subassemblies for shipment to new B-24 plants in Fort Worth and Tulsa. In record time, Sorensen's sketches on a San Diego hotel's stationery had become the largest single contract ever awarded by the US government. By April 1941, world events were driving demand for four-engine bombers and FDR issued orders for the United States to be able to build 500 four-engine bombers a month by June 1943. Ford was told to scrap its existing bomber subassembly plant plans. These rapid changes in the four-engine bomber production program would change the plans for building the bomber plant, for it was now going assemble flyable B-24 bombers. On June 2, 1941, production was elevated to 130 subassemblies per month and Ford was given a contract to produce 75 flyable B-24s per month at the Ypsilanti Bomber Plant. A bigger factory would be needed and the production floor space was now increased to 2,829,100 square feet. This plant would have a mile-long assembly line that was a quarter-mile wide. In October 1941, the Army elected to have even more operations completed in the bomber plant. To increase the plant size, Harry Hanson, Ford construction manager, decided to have the assembly line make a 90-degree turn inside the existing plant and extend to the south. This would keep the entire factory inside Washtenaw County to avoid any tax confusion with adjoining Wayne County should Ford elect to purchase the plant after the war. Until October 1941, the plant had been called the "Airplane Parts Manufacturing Building located near Ypsilanti." This caused confusion with a nearby Ford Ypsilanti electrical parts manufacturing facility: the new plant was then officially renamed the Willow Run Bomber Plant.

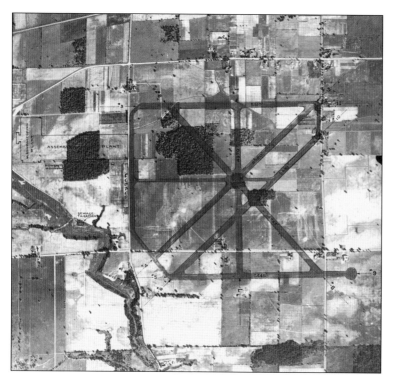

This aerial photograph shows the open farmland in western Wayne Country and eastern Washtenaw County, Michigan, where the bomber plant would be built. Sorensen had needed to quickly find three to four square miles of flat, open land upon which to build a factory and an airport. The factory would be built near the sleepy little creek shown running through the lower left of this photograph. That creek was known as Willow Run.

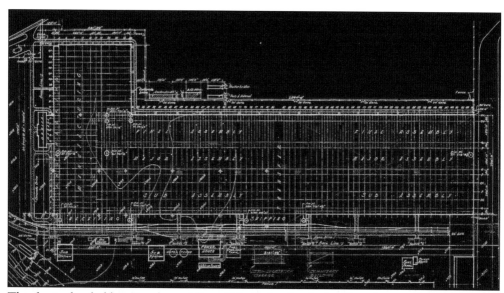

This facsimile of a blueprint shows Albert Kahn's plan for the bomber plant, as designed in June 1941. At this time, it was called the Airplane Parts Manufacturing Building and was located near Ypsilanti, Michigan. It does not include the famous 90-degree turn that was added by Ford to the east (right) side of the building in the fall of 1941. (BL.)

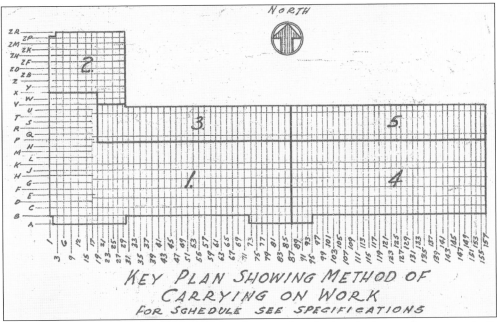

KEY PLAN SHOWING METHOD OF
CARRYING ON WORK
FOR SCHEDULE SEE SPECIFICATIONS

Shown here is the construction schedule for what was described as the largest factory ever built (up to that time). It was so big that one contractor could not handle the project. Five different construction contractors were given responsibly, each for different sections of the plant. Tight timetables were set with staggered starts for the five parts. Sections 1, 2, and 3 were for manufacturing and subassembly areas. Sections 4 and 5 were for final assembly areas that were added to the design in June 1941. (BL.)

This photograph, taken in June 1941, shows a detailed mock-up of the Willow Run plant with its mile-long assembly line, as designed by Albert Kahn and Associates. Originally, the manufacturing plant was going to be built at Ford's Rouge Complex and the subassembly plant near Ypsilanti. By early February 1941, it was elected to build the entire plant on Ford land near Ypsilanti. In June 1941, the plant design was expanded to include final assembly.

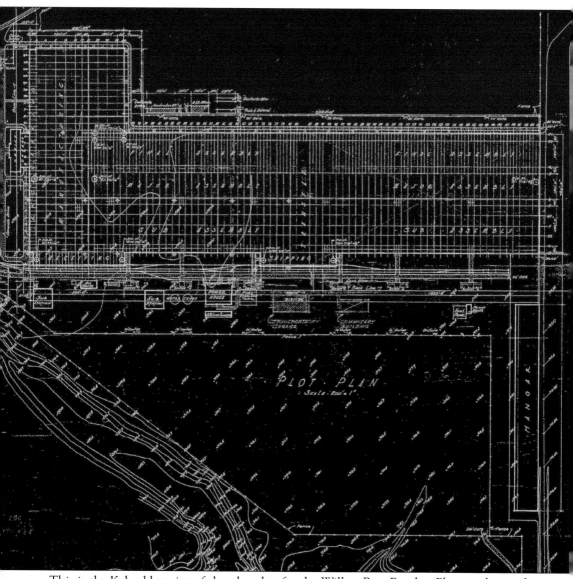

This is the Kahn blueprint of the plot plan for the Willow Run Bomber Plant as designed in June 1941, when Ford received a contract to build complete airplanes. When Ford got the flyable aircraft contract, the entire plant was redesigned by adding complete final assembly lines to the manufacturing and subassemblies areas. Under this plan, the bombers would emerge from the east-facing doors seen in the upper right of this blueprint. The east end of the bomber plant, on the right, is just inside the Washtenaw County border with Wayne County. The east-west locations in the factory are identified by frame numbers, starting with frame 1 at left and ending at frame 157 on the right. Note that Hangar One, identified as "Hangar," is in line with the end of the assembly plant and also right at the county line between Wayne and Washtenaw Counties. (BL.)

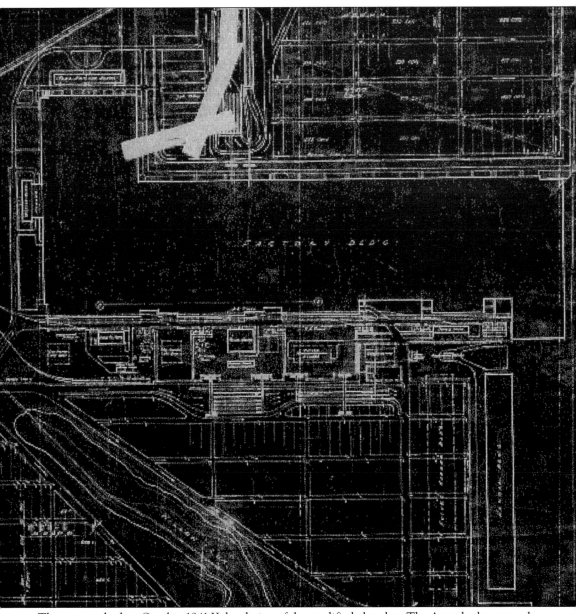

This is a patched-up October 1941 Kahn design of the modified plot plan. The Army had requested additional assembly operations in the bomber plant. This required an extension of the final assembly lines; however, due to its proximity to the county line, the plant could not be extended to the east (right) without crossing the border between Washtenaw County to the west and Wayne County to the east. Harry Hanson, Ford's plant construction manager, was concerned about the tax consequences of a cross-county factory following World War II, so he elected to turn the assembly lines 90 degrees to the south. The assembly line would then go through the existing manufacturing area at frame 141 and extend 200 feet beyond the existing southern wall. This would result in south-facing doors where bombers would emerge from the plant. This would also require Hangar One to be moved 320 feet to the west, allowing the bombers to exit the plant's new south-facing doors. (BL.)

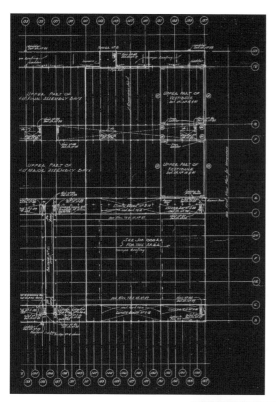

This is a blueprint of the east end of the bomber plant as designed in June 1941, showing the end of the long final assembly lines at the top of the photograph. Under this earlier plan, the bombers would have rolled out the east-facing doors shown at right. Shown in the lower portion of the photograph is part of the manufacturing area. (BL.)

This copy of a Kahn print from October 1941 shows a modification at the east end of the assembly line. Note the large "X" across the east end of the original bomber plan from frame 141 to frame 157 with the words, "See job notes for this area." Also of interest are the rough pencil sketches to the south extension of the assembly line and the area to the left of that, which would become the doping area for curing the fabric-control surfaces of the rudders, elevators, and ailerons. (BL.)

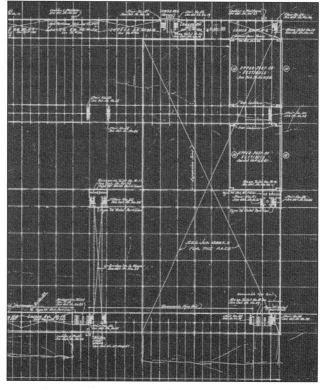

This Kahn print, dated October 19, 1941, shows the modified area of the final assembly lines with the 90-degree turn to the south and the 200-foot extension to the south-facing doors. In the center of the photograph is the 20-foot-wide center support structure that separated the north and south final assembly lines. The airplane-shaped forms in the middle of the print supported the elevators used to raise and lower the aircraft painters in the paint booths. (BL.)

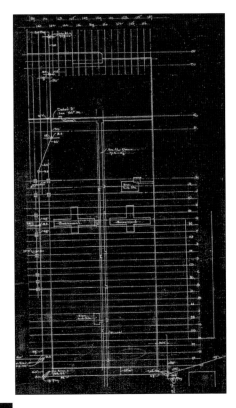

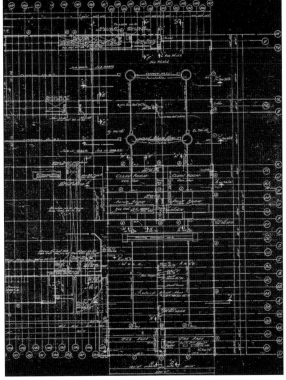

This Kahn print shows the finalized version of the assembly line extension and the four turntables that allowed the bombers to move into the "cleaning area" prior to being painted. In the center of these turntables is the unusual island-support structure needed to support the roof. The lines running down the middle of the assembly bays are the tracks where the bombers were pulled toward the south-facing doors. (BL.)

This north-facing view, taken in May 1941, shows that the steel structure for the north end of the manufacturing area in Section 1 had been started. Ground had been broken on April 18. The structure of two bays is clearly visible in this photograph. In the background on the left is the fruit and vegetable stand on Ecorse Road where the boys on the Ford farm used to sell their produce. (HF.)

This photograph, with a view looking east, shows plant construction in midsummer of 1941. At lower left, Section 1's structural steel has been erected. Shown off the end of the bomber plant are the stumps of the tress cleared for the huge facility. The center of the picture is where the future airfield was to be built, with trees having not yet been cleared from this area.

This photograph, with a view looking south in the late summer of 1941, shows that construction of Section 1 was well under way and construction of Section 2 had started. B-24 bomber assembly was the US government's top priority, so in the same month construction started on the Douglas plant in Tulsa, Oklahoma, and the Consolidated plant in Fort Worth, Texas. These plants together would receive 1,200 B-24 subassemblies from Willow Run.

Shown in this view looking east in the summer of 1941, the bomber plant is taking shape. The west end of Section 1, the manufacturing area, is almost complete at the lower right. Section 2, the tooling area, on the lower left is under construction. In the background, at left, Willow Run Airport has been laid out, and paving is already under way on the first east-west runway.

This photograph, taken in October 1941 with a view looking southeast, shows Section 1 complete, with Sections 2 and 3 still under construction. Parts for the bombers were already being produced in this part of the factory. In the foreground are Ecorse Road and the fruit and vegetable stand where the boys from the Ford farm used to sell their produce. In the background, at the top of the photograph, Willow Run Creek can be seen.

This photograph, taken in November 1941 with a view towards the southeast, shows structural steel of the North Final Assembly Bay in Section 5 being erected. In October 1941, it was decided to expand the bomber plant to include more final assembly operations in this building, which resulted in the famous 90-degree turntables in order to remain entirely within Washtenaw County. At that time, construction on this wall was temporarily put on hold at frame 139 due to pending changes in the plant design

This photograph, with a view looking northeast on November 28, 1941, shows the interior of Section 5, the nearly complete North Final Assembly Bay. At right is the original east-facing door where the bombers would have emerged from the assembly line. Turntables were now installed in front of these doors so completed airplanes could make a 90-degree turn to the south.

This photograph of Section 5's north wall, with a view looking northwest, shows the North Final Assembly Bay under construction. On the floor are some of the millions of wooden blocks that when laid would make up the floor of the assembly bays. The wood floor protected delicate parts and measurement equipment by absorbing impact energy if these items were accidentally dropped. The wood floor also absorbed vibration and noise created by operating machine tools nearby.

In January 1942, this photograph, with a view looking southwest, shows that construction of the original plant has been put on hold at frame 139 while a new support structure is being readied for building the assembly line extension to the south. In the background on the left, construction has started on Hangar One, with vertical steel uprights of the hangar bay's west wall visible. (BL.)

With a view looking north in February 1942, this image shows construction of the final assembly bay leading to the south-facing doors, where in April 1944, B-24s would roll out at the rate of one bomber per hour. The erected steel on the right is the center support structure that would divide the two parallel final assembly lines. To the left of the center support structure is the final assembly lines' west wall being erected.

In late February 1942, construction was quickly picking up; at left, the east wall of the East Final Assembly Bay has been put up. To the right of the crane is the unusual support-structure island that provided reinforcement for the roof over the turntables. At right is the North Final Assembly Bay where a temporary plywood wall had been built. Behind this plywood wall, the first Willow Run B-24 is being assembled. (BL.)

Hangar One, at that time known only as the "Hangar," is shown here in this February 1942 view that looks north from Bay 6 towards Bay 8. Hangar One was planned as part of the June 1941 bomber plant and would have been used for final assembly detail operations after the bombers were delivered out the original east-facing doors. In October 1941, the Army directed that these operations instead be completed inside the assembly plant.

Taken March 5, 1942, this photograph shows the final assembly bays leading to the south-facing delivery doors. On the left is the nearly complete west wall. On the right is the east wall showing only the vertically erected structural steel. Flying in the distance above the factory is a Consolidated-built B-24 coming into Willow Run for a landing.

Here, the Consolidated-built B-24 taxies by Hangar One. This airplane gave Ford a completed bomber to use in developing mass-production techniques for the Willow Run Ford-built B-24 bombers. To the rear of the B-24 is the east wall of the final assembly bays after the famous 90-degree turn was constructed.

Taken the same day as the previous image, this photograph shows a view facing west from the ramp. The Consolidated B-24 in front of Hangar One is still under construction. Bay 8 on the right (north) side shows that the hangar door had been installed by this time. Bay 8 now serves as the hangar for the Yankee Air Museum's flyable aircraft, including a B-17, a B-25, and a C-47.

This March 6, 1942, photograph shows the center of Hangar One, with the control tower on the left (south) side. There is a completed hangar door on Bay 5, and the Bay 6 door is partially complete. Note that this B-24 carries the US "meatball" insignia. This was removed from all US aircraft shortly thereafter because antiaircraft gunners would confuse it with the Japanese "meatball" found on a Japanese Zero.

This photograph with a view looking west on April 8, 1942, shows the Consolidated-built B-24 being pulled into Bay 8 at the north end of Hangar One, which was nearly complete at this time. To the rear on the right is the bomber plant still under construction, with the structural steel of the doping area visible. This illustrates how rapidly construction was progressing.

This photograph, taken less than a month later on April 20, 1942, with a view facing southwest, shows that the addition planned for the bomber plant in October 1941 is under construction a mere six months later. The bombers made their 90-degree rotations on turntables, seen on the right side of the picture. After the 90-degree turn, they entered the cleaning area and paint booth, then continued into the final finishing stations before rolling out the doors shown at far left.

Pictured on April 29, 1942, this view is facing north and shows the end of the assembly line at the south-facing East Final Assembly Bay. The top of the photograph shows the door structure supporting the 144-foot-wide door, where the bombers were rolled out. Even though the B-24 had only a 110-foot wingspan, the 144-foot-wide doors were required should the plant need to be converted later to build the larger B-29s or B-32s then on the drawing boards.

This April 21, 1942, photograph, with a view facing southwest, shows the final roof beam being installed at the end of the North Final Assembly Bay. These are the two original east-facing doors. The opening to the south of the east-facing doors can be seen at left where the cleaning area used prior to painting was located. In the rear, to the right of the crane, the center support structure is visible.

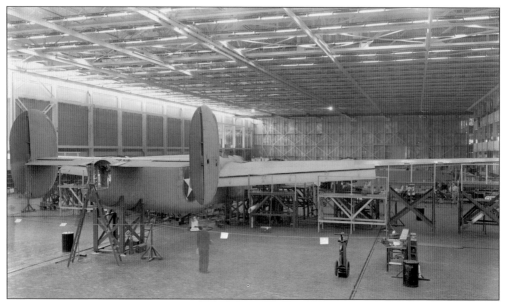

This May 12, 1942, photograph in the North Final Assembly Bay shows the first Ford-built aircraft contracted as an "educational order" for Ford engineers and workers to study in order to develop assembly operations. In the background, a temporary wall was built in October 1941 at frame 139 when the plant was redesigned. Beyond the plywood wall, the factory was still under construction. Some of the millions of wooden blocks that went into the floor can be seen.

This picture was taken in the East Final Assembly Bay with a view looking north from the south-facing door. It shows the barrier in the ceiling to separate the fueling stations from the rest of the factory. This is where the B-24s would be filled with 100-octane aviation gas before leaving the plant. This part of the bomber plant has been saved by the Michigan Aerospace Foundation and will become part of the Yankee Air Museum's future home.

This May 1942 photograph, with a view looking north, shows the nearly completed West Final Assembly Bay structure. The wall, with its opening for the B-24 wing and twin vertical stabilizers (seen in the back), isolated the paint booth from the rest of the factory. Through the center of the paint booth in the background is the island support structure located in the center of the turntables.

On May 15, 1942, the first Ford-built B-24 to emerge from Willow Run is shown running up its engines in front of the south-facing delivery doors. This airplane was a Ford aircraft built with Consolidated parts in order to fulfill an "Education Order" issued the year before. It was built to help teach Ford production employees how to build a B-24. The four 90-degree turntables allowed the original East and West Final Assembly Bay lines to feed into either of the new East or West Final Assembly Bays.

The finished bomber plant is shown in this photograph taken on June 6, 1942, with a view facing southeast. It shows a completed bomber plant and airport. To the right of the plant are the originally designed east-facing doors through which bombers were to be delivered. On the left are the south-facing doors where B-24 bombers were delivered two years later at the rate of one bomber an hour. The image shows the airport's runways: the east-west—9L-27R (left) and 9R-27L (right)—and northeast-southwest—5R-23L (upper center). These runways were all extended in the summer of 1943. The roads leading to the bomber plant on the left and lower right show the first two triple-decker overpasses built in the United States. From these "feeder roads," the highways linked the plant to Detroit, 25 miles to the east. These were built to facilitate getting workers to and from the factory, as there were no transit lines. In 1958, this would become part of the interstate highway system and be known as I-94.

This photograph, taken from the ramp in front of Hangar One, shows the bomber plant complex as of September 30, 1942. Hangar One is on the left with its airfield control tower at center. The bomber delivery doors are at the right. A B-26 twin-engine bomber is in front of the west door. Construction on the bomber plant was complete, but construction at Willow Run Airport would continue into 1944.

This photograph, taken in April 2016 from the identical perspective as the previous photograph, shows, other than modern jets in the picture, how little the ramp area in front of Hangar One has changed. At right are the south-facing doors from which the B-24s once emerged from the bomber plant. This area of the bomber plant has been preserved through the efforts of the Michigan Aerospace Association and the Yankee Air Museum.

Michigan's winter weather played havoc with Army airplane delivery schedules. Poor weather prevented doing engine run-ups and troubleshooting of mechanical problems discovered during inspections following release from the bomber plant. This photograph shows the run-up hangars under construction. These were later known as the Butler hangars and were demolished in 1998. It was from these hangars that Randy Hotton, a coauthor of this book, made his first solo airplane flight in a Cessna 172 on May 8, 1965. (BL.)

In the spring of 1943, Ford decided to build two hangars south of Hangar One for engine run-ups. This photograph shows a close-up of the rear of the completed run-up hangars. Note the louvers on the hangars, which allowed the airplane engines to be run while in the hangar, permitting engine testing even in foul weather. A B-24 is shown parked on the south side of the run-up hangars.

Taken in the summer of 1943, this photograph, with a view looking north, shows the bomber plant in full operation. The parking lots are full of workers' cars. Production is starting to accelerate. The ramp is covered with B-24s awaiting Army acceptance. A backlog had developed due to increasing deliveries because the facilities were not designed to handle the increased volume. The facilities would have to be expanded even more. At lower left are the nearly complete run-up hangars. The ramp extension to those hangars is still being built.

Taken in the early fall of 1943 with a view looking east, this photograph shows the completed run-up hangars and associated ramp area. At top, runways 9L-27R (left), 9R-27L (right), and 5R-23L (upper center) have all been extended. In the foreground, a material supply warehouse is under construction. This warehouse served as both a supply center for the manufacturing plant and a distribution center for completed B-24 replacement parts needed all over the world.

This image with a view looking northeast in the early fall of 1943 shows the structure officially known as the Spare Parts, Stock Storage, Receiving and Shipping Building under construction. The twin water towers of the bomber plant are visible in the right background. This building would later become part of a General Motors auto-assembly plant, where in 1959 the Chevrolet Corvair would be built. (BL.)

This photograph, also taken in early fall of 1943, and with a view looking north, shows the ground breaking and start of construction for Hangar Two. Hangar Two would be nearly identical to Hangar One. Additional land south of the airport had been purchased in the summer of 1943. Ramps and taxiways were being extended to the south to support the new hangar.

This photograph, taken on November 11, 1943, shows Hangar Two nearing completion. There were many changes in the design and manufacturing of the B-24 between 1941 and 1945. Hangar Two would allow additional space for aircraft modifications. More hangar space was badly needed in order to meet the Army's demanding delivery schedule that had grown beyond Hangar One's capacity. This hangar also served as the Yankee Air Museum's temporary home after its original World War II hangar burned down in 2004. (BL.)

Taken in the late December 1943, this photograph shows the completed Hangar Two, located at the southwest corner of the airport. When the bomber plant was designed in 1941, only 75 flyable airplanes were scheduled to be built per month. By the winter of 1944, flyable B-24s were coming off the Willow Run assembly line at the rate of nearly 400 a month. This hangar was taken down in 2015. (BL.)

This winter 1944 photograph from the inside of Hangar Two, with a view looking south from the top of the Bay 8, shows how quickly this hangar space was put to work. Shown in this picture are 16 B-24Hs-Block 15s in various stages of modification. This is the final production run of USAAF Contract No. 21216-5, issued in August 1942. Its serial number is 4294970. These are among the last of the painted Ford B-24s before the switch to the bare aluminum surfaces for the remainder of World War II. Note how the US insignia differs from that seen in 1942.

This photograph from the summer of 1945 shows a single-tail B-24 flying over the Willow Run Bomber Plant and airport. In March 1945, Ford received Contract No. 21216-47 for 2,175 of the single-tail B-24N, but this order was cancelled as the war in Europe came to a conclusion. The completed Hangar Two is located at lower left. The ramp is covered with war-weary B-24 bombers from the European theater. These B-24s were flown into Willow Run to be reconditioned and updated for service in the Pacific theater. The rapid conclusion to the war with Japan ended the need for additional B-24s in the Pacific, and these airplanes were flown to reclamation centers where they were scrapped.

Seven

OBSTACLES TO PRODUCTION

Ford's B-24 production plan was so beset with obstacles it seemed like it would be impossible to reach the one-bomber-per-hour production goal. By spring 1942, the lack of production at Willow Run had gained public attention. The first production airplane was not delivered until October 1942, and Willow Run was being nicknamed "Will it run?" It would be a steep learning curve for Ford. Sorensen found Consolidated had no B-24 blueprints. This lack of design documentation dashed his original plan of quickly making the tools to ramp up production. Ford sent engineers to San Diego, and they produced 30,000 drawings, of which many would be obsolete when they reached Detroit. Lacking blueprints was a major setback for B-24 mass production. The initial Ford B-24 production plan rested upon a major assumption that the B-24 design would not change. The Army's constantly mandated changes, such as rerouting of fuel lines, regularly shut down production. The design needed to be frozen, and Sorensen convinced General Arnold that changes should be incorporated at certain planned intervals, like automotive annual model changes This was the foundation of the block concept of aircraft production and was adopted by the entire aircraft industry. The lack of a trained workforce inhibited full production. Hiring new workers was put on hold in late 1942 until new on-the-job employees were trained, but turnover delayed building a trained workforce. Much of the employee turnover was caused by deplorable living conditions near Willow Run; the plan to get a stable workforce was hobbled by the lack of housing. The government acquired additional land to the east of Ypsilanti from Henry Ford and built housing to be used by the Willow Run employees. On March 12, 1943, shortly before his death in May 1943, Edsel Ford appointed veteran Ford employee Mead Bricker as Willow Run's general manager. Bricker's 505 program would be the solution to production delays by outsourcing manufacturing to support Willow Run assembly operations. Gaining momentum, in October 1943, Willow Run produced 365 bombers. New production schedules were drawn up, and in December 1943, more subassembly work was shifted to outside plants. Despite all the obstacles, under Bricker's guidance, Willow Run came through and would set B-24 production records. In April 1944, Willow Run finally reached Sorensen's goal of building one bomber an hour.

In the fall of 1942, production was lagging and continued employee turnover was putting tremendous pressure on Willow Run. Ford had long experience in making productive workers, and during the war, it did teach factory skills to thousands of new employees. Unfortunately, a high percentage of the workers took what they learned in Willow Run's training programs and immediately secured work in other Detroit plants near adequate housing, which could not be found in the Willow Run area. (HF.)

The turnover at Willow Run prevented building up a trained workforce. When production ended in 1945, a total of 80,774 employees had been hired at Willow Run. Of these, only 30,021 stayed more than one year, 34,533 stayed less than three months, and over 12,000 quit after just 10 days. For example, in January 1943, Willow Run hired 1,186 new employees but lost 1,669.

One of the things driving a great deal of this turnover was the deplorable conditions faced by the workers flooding into the Willow Run area from Appalachia and elsewhere in the South. There were stories of workers living in chicken coops, trailers, shanties, and even in their own cars. Adding to the misery was a complete lack of infrastructure like water, sewer, and public services. Even schools were inadequate to handle the influx, and it was hard to find teachers. This combined with men leaving for military service drove a terrific rate of employee turnover.

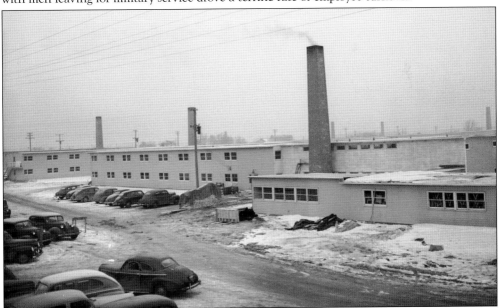

In early 1943, to ease the housing crisis, a dormitory was built adjacent to the west end of the bomber plant. The building provided rooms for single persons. The conditions there could be called spartan at best, with paper-thin walls, communal bathroom facilities, and the 24-hour comings and goings of bomber plant employees and their guests.

Ford Motor Company
Bomber Plant
Willow Run - In Detroit Area
WILL EMPLOY
Men and Women

1. To Train for Work in Aircraft Industry, or
2. Those Who Already Have Experience or Training.

Work In World's Largest Bomber Plant.

Essential Industry
Excellent Opportunity for Advancement
Women Paid Same Hourly Rate as Men
Modern, New Buildings.   Clean Work, Interesting
and Pleasant
54-Hour Week. 1½ Pay for Over 40 Hours
Employer will Pay Transportation

REQUIRES: Age 18 or Older.   Draft Deferred.
Physical Examination

Those now Employed in War Industry or Farm Work will not
be Considered Unless Eligible for Statement of Availability
Under W. M. C. Regulations.

APPLY IN PERSON
Monday or Tuesday
NOVEMBER 8   -:-   NOVEMBER 9
United States Employment Service
SOMERSET   Fountain Square   KENTUCKY

 PATRIOTISM: It is urgent that Men and Women be
Procured to fill these Jobs.   If you are already employed full time, consider it your duty to place this Advertisement in the hands of a person who might be In-
terested.

In an effort to fill open positions at Willow Run, Ford officials fanned out to towns throughout the upper Midwest and even into the hollows of Appalachia. Ford placed this recruiting advertisement in Kentucky in November 1943. It offers employment to both men and women at equal wages. The wages paid at Willow Run were often 10 times the prevailing wages in places like Somerset, Kentucky. Rose Monroe, the prototype for Rosie the Riveter, was one of the women from Somerset. (Timothy O'Callaghan.)

In 1943, Henry Ford sold more of his remaining farmland north of Michigan Avenue, and plans were made to build 2,500 units of temporary housing. Shown here is the federal housing community called Willow Village, which opened in December 1943. By that time, major outsourcing of B-24 parts was well under way, reducing total employment and the need for additional housing. The housing in Willow Village was used by World War II veterans following the war while they attended school on the GI Bill.

Shown here are draftsmen producing B-24 blueprints. Sorensen had asked Consolidated for a full set of blueprints, but the company did not have complete sets of up-to-date ones. It only had out-of-date blueprints, templates, drawings, and sketches. In March 1941, Ford had to send 200 engineers and draftsman to San Diego. They worked three 8-hour shifts a day, working seven days a week, to produce five miles of engineering drawings a day. They spent nearly four months in San Diego producing engineering drawings. The Army's demands for constant design changes meant that 10,000 of the drawings would be obsolete by the time that airplane flew. The learning curve for Ford Motor Company became very steep.

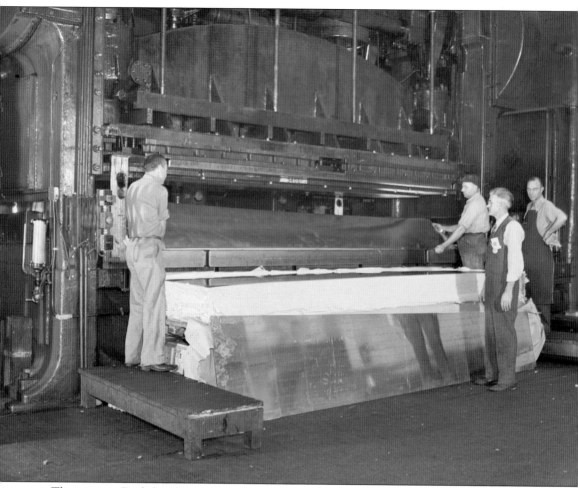

The secret to Ford's bomber-an-hour plan was use of hard steel dies in stamping presses, as shown here, to turn out standardized aluminum parts, contrary to aircraft industry use of soft kirksite dies. Ford decided to use the hard metal dies shown here because, for over 35 years, they had been used to produce volumes of standardized automotive and truck parts. Once set up, tested, and put into action in high-speed presses, these dies, manned by relatively unskilled employees, could turn out extremely accurate parts in very little time. Tests proved that the strength of the aluminum was not significantly impaired by the use of hard dies. Ford had concluded that maintenance costs and time lost to repair and service soft dies would slow down production.

The whole Ford B-24 production plan rested upon a major assumption that the B-24 design would be reasonably stable. However, the Army Air Corps was used to airplane design being modified when reports came in from the field. This inability to stabilize designs prevented Ford from rapid manufacturing of identical airplanes until late 1942. For example, in the first year of B-24 production, the Army ordered 575 master changes. Every master change resulted in assembly line shutdown and massive reengineering.

The block concept of aircraft production became the answer to constant design changes. The block concept allowed the design to be frozen for a specific number of aircraft. A B-24J-10 (Block 10), after being equipped with a new oxygen system, became a B-24J-15 (Block 15). In order to make changes to airplanes without shutting down the assembly, modification centers were opened where updates could be incorporated into the previously built airplanes. Shown here is Hanger Two at Willow Run, where B-24s were modified.

Six months lost in development of tools and dies combined with constant design changes resulted, at first, in no B-24s being completed on schedule. Everyone was wondering when the B-24s would reach combat units. The Willow Run plant was nicknamed "Will It Run?" The Truman Committee, headed by Sen. Harry Truman (second from left), shown here with Charles Sorensen (third from left), was formed to investigate possible malfeasance at Willow Run and other war plants. The committee left with the pronouncement that Willow Run "compares favorably" with any other airplane plant in the country.

In March 1943, Mead Bricker (third from left) took over as general manager of Willow Run for Ford. He is shown here with Edsel (second from left) and Henry Ford (right). Bricker became the authority for Willow Run's bomber production. Since it was impossible to retain a workforce that would support the ambitious production goals of 1941, outsourcing of manufacturing parts and subassemblies to support the bomber assembly plant would be the answer. Under Bricker's guidance, Willow Run then proceeded to set B-24 production records.

Eight

ROSIE THE RIVETER

A major problem for Ford and other manufacturers during World War II was employee turnover caused by so many men entering military service. By 1944, there were 15 million men in the US armed forces. The manpower shortage was so severe that it was impossible to fill all the open jobs. In the aircraft industry, this shortage was met by women. For many Americans, nothing symbolized national effort on the home front more than women working in aircraft plants, as symbolized by Rosie the Riveter. This would be a breakthrough in airplane manufacturing. At first, companies had doubts about hiring women. However, experience soon showed that women could be especially skilled in subassemblies. Lightweight tasks, such as electrical wiring bundles, were areas where women excelled. In World War II, women filled positions that had previously been defined as "male only" jobs. An underlying advantage was that, when a woman replaced a man, she tended to stay in that job. When an effort was undertaken to find a real worker to represent Rosie, she was discovered by an actor from Hollywood named Walter Pidgeon. Her name was Rose Monroe, a widow from Kentucky. Monroe was helping build the B-24 that would crush the Axis powers, and thus became the iconic Rosie the Riveter. Although the women at Willow Run earned the same wages as men, the "Rosies" were motivated by much more than money. They could not serve overseas, but they had brothers, sons, husbands, and fathers fighting for them. They wanted to do their parts to help those men. In 1945, total employment in the auto industry was 800,000, and women filled two-thirds of the work positions. Without women workers, which made up nearly 40 percent of the total workforce at the end of the war, bomber production goals would never have been met. In a time of national crisis, American women stepped up to meet the challenge. With quiet confidence that exceeded expectations, they produced a deluge of bombers and other material. Americans must never forget what these women did as they boldly faced the task of building the vast armada of bombers, tanks, trucks, guns, and ships.

The government did not have quotas for hiring women but encouraged them to work in war production due to the shortage of manpower. To do this, the government distributed a poster called *We Can Do It!* It shows a woman in denim overalls with a red polka-dotted bandana. This became the classic "Rosie" persona. There was no specific model for this woman.

A major problem for Ford was the high turnover rate of employees. This shortage was driven by a scarcity of acceptable housing plus a turnover of men, who were drafted or volunteered for military service. A solution to the manpower shortage would be womanpower, shown here as women start their shifts at Willow Run. It became a breakthrough in airplane manufacturing. For many Americans, nothing symbolized the national effort on the home front more than women working in aircraft factories.

About the same time that the airplane *Rosie* came on the scene, Norman Rockwell created his famous *Saturday Evening Post* magazine cover. The magazine cover of *Rosie the Riveter*, a burly woman wielding a heavy-duty shipbuilding rivet gun, is not a true depiction of the working women in the factories, such as the one shown here. The true "Rosie" was an average, 110-pound woman using a lightweight rivet gun and assembling airplane parts.

Shown here is a woman stitching fabric on a B-24 elevator at Willow Run. The control surfaces on the B-24 were covered in fabric. She is filling the traditional role then expected of a woman—sewing. Time would prove that women were capable of much more than stereotypical "women's jobs."

The job that gave the title to Rosie the Riveter was driving rivets. B-24 bombers contained more than 300,000 rivets of 500 different sizes used to fastened aluminum body parts together. Shown here are women driving rivets into a B-24 center wing. The center wing was in held place by 27-ton fixtures designed to ensure that tolerances of .002 were maintained. Here, the women are standing on a platform that rose and fell to ease the job of driving the rivets, which was the most labor-intensive task in bomber assembly. Building a bomber an hour meant that over 300,000 rivets were being set every hour, one at time, by hand. (HF.)

Women's unique skills in light work were as vital to production in aviation as the heavy-lifting tasks. Light tasks, such as electrical circuit work, including threading, running, connecting, and checking electrical connections, were areas where women excelled and even outproduced their male counterparts. One company started putting women to work in tubing departments and saw production jump by 20 percent. Soon, other companies started hiring women for every department.

Shown here is a woman operating an overhead crane in the Willow Run Bomber Plant. The factory was designed with hundreds of overhead cranes, with capacities to hold up to 20,000 pounds, that could move anything to anywhere within the factory. These cranes were vital to Willow Run's production flow, and almost all were operated by women.

The women filled many jobs that had previously been defined as "male only," such as operating lathes. An underlying advantage was that when a woman replaced a man, she tended to stay in that job until the end of the war. As a result, the percentage of women working in the aircraft industry continued to grow until the war's end.

Shown here is a woman inspector examining a part before it advances to assembly operations. Women at Willow Run earned the same hourly wages as men in the same job classification. Ford also rapidly moved Willow Run women from production jobs to supervisory roles. These women were motivated by more than pay. They could not fight overseas, but they wanted to do their shares to help the men overseas by making sure that all those parts were perfectly made.

Shown here is a photograph of Rose Monroe, a widow from Somerset, Kentucky. She had traveled from Kentucky to Michigan with her two daughters to work in the bomber plant. Rose had two daughters, who stayed with friends at a farm near Jackson, Michigan, 50 miles west of Willow Run. They would spend the weekends with their mother in her dormitory room at Willow Run. Rose wanted to be a pilot and after leaving Willow Run would earn her pilot's licence. (Vicky Croston.)

Shown here autographing a B-24 is Walter Pidgeon, a famous Hollywood actor. He was touring the plant when he was introduced to an attractive riveter with curly brown hair and assembly tools in her hands. "She was perfect!" Pidgeon recalled, "Here was a real honest-to-God, hardworking lady whose name was Rosie!" Thus, Rose Monroe became the iconic Rosie the Riveter.

Here, on June, 28, 1945, "Rosies" surround the last Ford-built B-24 as it rolls off the Willow Run assembly line. In 1945, women aircraft workers made up nearly 40 percent of the US workforce. In a time of national crisis, these women stepped up to meet the challenge. There was no one else to do the job if the women had not stepped up. As Donald Miller writes in his book *Masters of the Air*, World War II was a bomber war. It has been said that the massive application of heavy bombers brought the war to a quick end and saved the lives of tens of thousands in the Allied military. America could never have had the bomber force needed to quickly end the war without the Rosies. They made a difference that must be celebrated and remembered for generations to come.

Nine

LINDBERGH
AND DOOLITTLE

The iconic pilot Charles Lindbergh, known for his solo single-engine flight across the Atlantic in 1927, became deeply involved in the perfection of B-24 production at Willow Run. Due to the fact "Lucky Lindy" was active in the isolationist America First Movement before World War II, Roosevelt had ordered that Lindbergh be treated as an outcast from the aircraft industry. Due to his close relationship with Henry Ford, though, Lindbergh asked if he could work for Ford. Ford asked the War Department if he could hire Lindbergh to help Ford's aircraft production, and the department gave its approval. By virtue of this ironic twist of fortune in Lindbergh's life, he ended up in Detroit and brought desperately needed knowledge of aircraft construction to Willow Run. What he found in spring 1942 at Willow Run was appalling to a man who understood aircraft design. Willow Run, now in the spotlight, initially was producing parts that were lower quality than even the aircraft industry had predicted. As the workforce stabilized, production quality at Willow Run rapidly improved. When the famous pilot departed from Willow Run in the spring 1944, he made this entry in his diary: "Took a bomber up for initial shakedown flight. Plane in excellent condition so ran Army acceptance flight procedure also."

Jimmy Doolittle, famed for his daring raid on Tokyo by Army bombers from a US Navy aircraft carrier in April 1942, played a major role in the development of the Willow Run Bomber Plant. Known for setting aircraft performance records in the 1920s and 1930s, Doolittle earned both his master's and doctoral degrees in aeronautics from the Massachusetts Institute of Technology. He had been recalled to active duty from the reserves by an act of Congress on July 1, 1940, and was stationed in Detroit as a liaison officer to assist the auto industry. His role, as he described it, was to perform a "shotgun marriage between the auto industry and the aircraft industry." He remained assigned to Detroit until January 1942. At Knudsen's meeting with auto companies in Detroit, the Air Corps men were led by Doolittle. In December 1940, he visited Henry Ford to see if Ford would build B-24 parts. This would lead to Sorensen's visit to San Diego and his proposal to build one complete bomber an hour. Doolittle had a keen interest in Ford's bomber production.

Lindbergh is shown here with the *Spirit of St. Louis* (Registration: N-X-211). It was in this custom-built, single-engine, single-seat monoplane that Lindbergh flew solo across the Atlantic Ocean between May 20 and 21, 1927. Lindbergh took off in the *Spirit of St. Louis* from Roosevelt Airfield in Garden City (Long Island), New York, and landed 33 hours, 30 minutes later at Aéroport Le Bourget in Paris, France, a distance of approximately 3,600 miles. This flight made Lindbergh the most famous aviator in the world and an instant celebrity. (LC.)

Shown here in the summer of 1927 were Charles Lindbergh and Henry Ford standing next to the *Spirit of St Louis*, the airplane Lindbergh had flown across the Atlantic Ocean. Lindbergh would give Henry Ford his first and only airplane ride in it at a stop at the Ford Airport in Dearborn while making his national tour following his transatlantic flight. Lindbergh's Ryan monoplane was custom-built in San Diego more than a decade before the first B-24. (HF.)

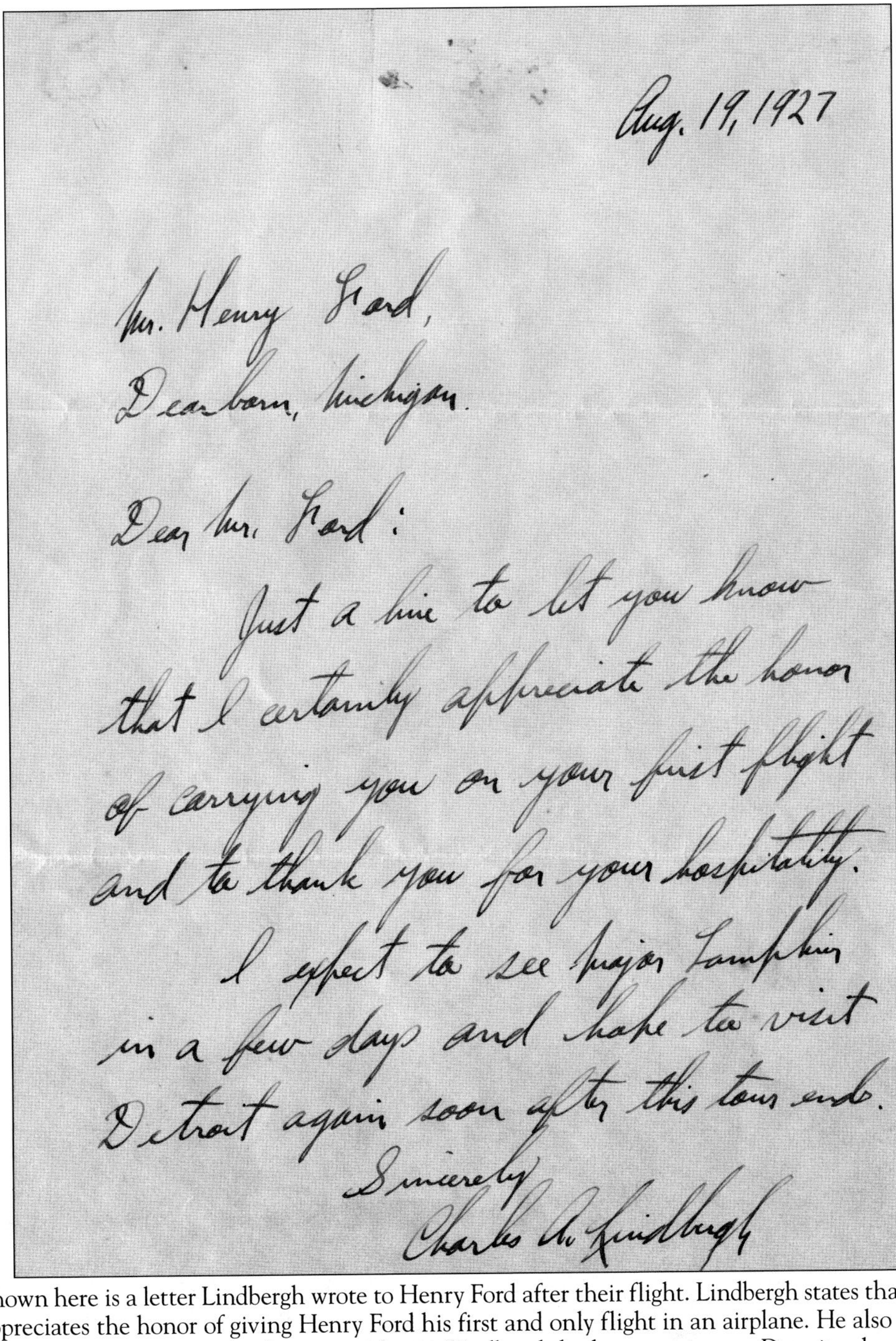

Aug. 19, 1927

Mr. Henry Ford,
Dearborn, Michigan.

Dear Mr. Ford:

Just a line to let you know that I certainly appreciate the honor of carrying you on your first flight and to thank you for your hospitality.

I expect to see Major Lampkin in a few days and hope to visit Detroit again soon after this tour ends.

Sincerely
Charles A. Lindbergh

Shown here is a letter Lindbergh wrote to Henry Ford after their flight. Lindbergh states that he appreciates the honor of giving Henry Ford his first and only flight in an airplane. He also says he hopes to visit again after his national tour. Lindbergh had connections to Detroit, where he was born and where his mother still lived and worked as a schoolteacher. He would visit Henry Ford at the latter's Fair Lane home many times during the 1930s, and the two developed a close relationship. Both Ford and Lindbergh had received medals from Germany, and as the war grew closer, both were branded as Nazi sympathizers by the Roosevelt administration. This close relationship would bring Lindbergh to Willow Run during World War II.

Here is Lindbergh, on the right, when he came to Willow Run in the spring of 1942; at far left is Charles Sorensen. What Lindbergh found was appalling to a man who understood aircraft design. When looking over a knockdown part to be shipped to Fort Worth and Tulsa, he was quoted as saying, "It was the worse piece of aircraft construction I have ever seen. . . . Inexperience shows everywhere."

This image from the summer of 1942 shows Lindbergh in Willow Run's Hangar One. Willow Run was in the spotlight because production was lagging. Only one percent of the bomber plant's workers had any experience building airplanes. Lindbergh took a stance against Sorensen and Edsel Ford, who wanted to hire more inexperienced employees. He sided with Mead Bricker, who felt production could accelerate faster if they spent more time training their present employees. Time and accelerating production would prove that more training was the solution to lagging production.

Jimmy Doolittle, in addition to having a PhD in aeronautical engineering, was a skilled aviator and technical expert who made a name for himself in the 1930s as one of the nation's leading air race pilots. He won the Thompson Trophy Race in 1932 with an average speed of 252.7 miles per hour. His work with Shell Oil Company to make 100-octane aviation gasoline, which gave the Allies a tremendous advantage during World War II. Major Doolittle was recalled to active duty by an act of Congress and became involved with Bill Knudsen to figure out how to get the auto industry to build four-engine bombers in high volume. To do this, the Army ordered him to Detroit as the assistant district supervisor of the Central Air Corps Procurement District. On January 9, 1942, Doolittle was ordered to Headquarters Army Air Forces in Washington, DC, where he was tasked with leading his famous Tokyo air raid from the deck of the Navy's aircraft carrier *Hornet*. (HF.)

On October 23, 1941, Jimmy Doolittle made the first official landing at Willow Run Airport in a Howard DGA. The DGA was a fast four-seat racing aircraft that had won the Bendix and Thompson Trophies in 1935. Although this was the first "official" landing, other aircraft had landed at Willow Run Airport in the previous days.

Shown in this picture is Jimmy Doolittle (right) in civilian clothes beside his civilian aircraft, along with Charles Sorensen (center). When Doolittle was assigned to Detroit, he was given an office at Ford Headquarters on Schaefer Road in Dearborn. This Army Air Corps post gave him complete authority to coordinate all activities of automotive manufacturing concerns on the conversion of automotive plants to airplane parts manufacturing. He worked closely with Sorensen and played a major role in bringing Ford into building B-24s.

After leaving Willow Run, Jimmy Doolittle became involved in planning a raid on Japan. This was to be known as the "Tokyo Raid," but it would later become known as the "Doolittle Raid." On Saturday, April 18, 1942, a B-25 is leaving the deck of a Navy aircraft carrier, the USS *Hornet* (CV-8), on its way to Japan. Sixteen US Army Air Force twin-engine B-25 bombers were launched deep in the western Pacific Ocean; the plan called for them to bomb military targets in Japan and then continue westward to land in China. Fifteen aircraft reached China but all crashed, while the 16th landed at Vladivostok in the Soviet Union. The raid caused negligible material damage to Japan, but it succeeded in its goal of raising American morale and casting doubt in Japan on the ability of its military leaders to defend their home islands. (Doolittle Raiders.)

In December 1940, Jimmy Doolittle had approached Henry Ford about building the B-24. Doolittle had been given complete charge of getting the auto industry involved in aircraft and engine production and had a keen interest in the proposed Willow Run Bomber Plant. After the April 1942 raid on Japan, Doolittle (fifth from right) paid a visit to Willow Run on May 29, 1942; here, he is posing with Henry Ford (in hat and suit) below the nose of the first B-24 built at Willow Run. Second from left is Charles Lindbergh in a dark suit, and to the immediate right of Doolittle are Edsel Ford and Charles Sorensen. Doolittle initially believed that loss of all his aircraft would lead to his being court-martialed, but for his leadership on the Tokyo Raid, Doolittle was promoted to general and awarded the Congressional Medal of Honor. Eighteen months later, he took command of the Eighth Air Force, flying out of England. There, he would issue orders for the Ford-built B-24s to fly to their targets in German-occupied Europe.

Ten

A Bomber an Hour

In the spring of 1944, Willow Run had overcome its many obstacles and reached Sorensen's bomber-an-hour goal. When the Army heard that Ford had said he could build a bomber an hour, officials said he had pulled this number out of his hat. The government and military leaders did not understand the miracles that could be pulled off by Detroit's skilled production people. To obtain this production miracle, Ford had broken the fuselage design into 33 different sections. Precision tooling was developed that held tolerances to Ford's standard 1/1,000 of an inch. This made all the parts interchangeable. Average time from when a part was stamped until a bomber rolled out the bomber plant doors was 25 days. The heart of B-24 production was the center wing assembly, and this was where Ford's tooling experience showed itself to be far superior to traditional airplane-makers. The South Subassembly Bay was where the center wing was built. The wings were then moved into horizontal one-hour stations where bomber racks and turbo chargers were installed. Parallel to the South 150' Center Wing Subassembly Bay was the North 150' Fuselage Subassembly Bay. This is where the front and rear fuselage subassemblies were manufactured. Center wings were placed on four separate assembly lines in what was known as "four-hour" stations. These assemblies would move forward every four hours during assembly operations. With the center wing and fuselage section subassemblies completed, the bomber moved into final assembly. Production now moved rapidly to deliver a completed B-24 bomber in seven days. In the spring of 1944, the auto industry, in its many contributions, was building 50 percent of four-engine bomber airframe weight and 75 percent of the bomber engines. There would not have been a bomber force capable of quickly ending the war without the auto industry. Following World War II, the Army evaluated the auto industry's contribution to aircraft production and said that "the question of whether the automotive industry could be converted to production of aircraft was settled beyond debate during the war. The success left no room for argument. Given the limited capacity of the aircraft industry and the enormous pressure for plants after May 1940, it is fair to conclude that the AAF was justified in underwriting the Ford experiment." In 1944, Willow Run would produce 92,568,000 pounds of airframe. The plant was the manifestation of the miracle of the Detroit's manufacturing mastery.

FORD B-24 BREAKDOWN

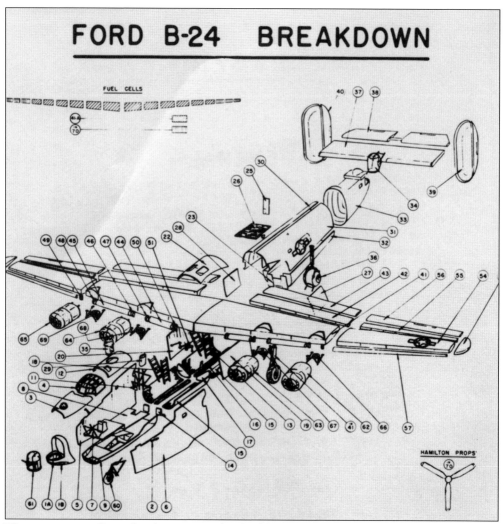

This is an original item list for parts used to build a B-24 bomber. When Ford was authorized to build B-24 subassemblies on February 21, 1941, it had to coordinate with airplane developer Consolidated Aircraft on component identification. Consolidated's specification Z1-32-007 gave Ford a list of items used in building the B-24, with items numbered from 1 to 70. This material list was to be used by Ford to develop manufacturing processes. Samples of these parts were to be shipped from Consolidated to build B-24 subassemblies, but they were not shipped until July 1941. By that time, Ford had been ordered by the Army to build complete airplanes. The item numbering system was carried forward and expanded for the assembly of complete airplanes and used the entire time B-24s were assembled at Willow Run. (HF.)

DEPT.	ITEM	PART NUMBER	NAME
932F	1A	32B 9550-0 &	Nose Enclosure
P.F.	1B	32B 9532	Plexiglas Nose
916H	2	32F 3135-3L/R	Pitot Tubes
938A & P.F.	3	32B 1705-2	Fuse. Nose Sect. Upper Frt 0.1 to 1.2
P.F.	4	32B 9144	Pilot's Enclosure
938	5	32B 1737-6	Fuse. Nose Side Panel R.H. 0.1 to 4.15
938B	6	32B 1736-2	Fuse. Nose Side Panel L.H. 0.1 to 4.15
914	7	32B 1725-2	Pilot's Floor
914-942A	8	32B 1735-2	Radio Operator's Floor
938A	9	32B 9800	Fuse. Nose Bottom Panel Sta. 0.1 to 4.0
	10		Cancelled
942A	11	32B 1741-0	Truss Blkhd. Sta. 4.1
912	12	32B 1727-2	Nose Sec. Upper Rear Deck 2.0 to 4.2
P.F.	13	32F 4799-2-3-4-5	Bomb Rack
916 & P.F.	14	32B 9145-L/R	-32L/R Segments Blkhd. Sta 4.0 Segments
P.F.	15	32F 046	(-139-L/R) (-140-L/R) Bomb Bay Doors
P.F.	16	32B 9014-0	Lower Longeron 4.0 to 6.0
942 & P.F.	17	32B 1750-0L	Blkhd. 5.0 L.H. Portion
942 & P.F.	18	32B 1750-0R	Blkhd. 5.0 R.H. Portion
938A	19	32B 1795-2	Side Panel L.H. 4.15 to 5.25
938A	20	32B 1796-2	Side Panel R.H. 4.15 to 5.25
937	21	32F 9038	Truss Rear Bomb Rack
912	22	32B 1708-2	Fuse. Top Deck Abv. Wing 4.2 to 5.1
942	23	32B 9148-L/R	Segment Blkhd. 6.0
	24		Cancelled
P.F.	25	32B 11218-5	Door Blkhd. 6.0
942	26	32B 1755-2	Floor 5.1 to 6.0
938A	27	32B 1784-2	Fuse. Side Panel Below Wing 5.25 to 6.0 L.H.
938A	28	32B 1785-2	Fuse. Side Panel Below Wing 5.25 to 6.0 R.H.
P.F.	29	32F 9712	Hydraulic Reservoir Tank
939	30	32B 1788	Fuse. Upper R.H. Sect. 5.1 to 7.7
939	31	32B 1787	Fuse. Upper L.H. Sect. 5.1 to 7.7
939A	32	32B 1786-0	Fuse. Bottom Section 6.0 to 7.7
939-940A-940H	33	32B 1794-0	Fuse. Tail Section Aft. Sta. 7.7
P.F.	34	32F 22616	Motor Products Tail Turret
GFE	35	32GF8227-6A3-0	G.L. Martin Elec. Power Driven Turret SK6276
GFE	36	32F 8673	Sperry Turret
945-912	37	32T 9352	Stabilizer
912-946-915C	38	32T 10503-2L/R	Elevator Assembly
912-915	39	32T 8050-0	Fins
912-946-915C	40	32T 10115-3L/R	Rudders
936	41	32W 12002-4	Wing Center Section Vertical
937	41	32W 1701-P	Wing Center Section Horiz. (Incl. Main L. G.)
P.F.	41A	32G 1039-4	12 Fuel Cells Center Wing. Plus 6 cells
934-931	41B	32W 10159-3L/R	Anti-icing Fluid Tank & Wheel Fairing
P.F.	42	32W 520-2L/R	Wing Center Section Trailing Edge
P.F.	42	32W 935C-L/R	Short Section-Trailing Edge
P.F.	43	32W 500-2L/R	Flap
943A	44	32W 1702-6L/R	Center Wing Section Leading Edge
943A	45	32W 1702-7L/R	Wing Center Sec. Leading Edge Between Nac.
943A	46	32W 2067-2L/R	Inbd. Nacelle to Wing L.E. Fairing Inbd.
943A	47	32W 2068-2L/R	Inbd. Nacelle to Wing L.E. Fairing Outbd.
943A	48	32W 2069-L/R	Nacelle L.E. Attaching Stub
935A	49	32W 2070-L/R	Nacelle L.E. Connection Stub & Access Panel
943A	50	32W 302-01/R	Wing Center Sec. Leading Edge Attach Panel
932	51	Misc.	Truss Fuse. to Wing 4.1 to 4.2 Item 51 is a group of loose parts installed at Final Assy.
	52		Cancelled
947	53	Misc.	Flap Track Supports. Group of loose parts. Installed at Final Assembly
935H	54	32X 12003-2L/R	Wing Outer Panel
P.F.	55	32X 026-L/R	Wing Outer Panel Trailing Edge
912-946-915C	56	32W 10573-3L/R	Aileron
935A	57	32W 9392-1/R	Wing Outer Panel Leading Edge
P.F.	58	32W 1024-0L/R	Wing Tip
	59		Combined with Item 32
937-906	60	32W 8879 32L 9141 32L 9129 32L 9147	Nose Landing Gear
GFE	61	32G 8687	Emerson Turret
919	62	32P 9101	Nac. Power Plant Inst. L.H. Outbd.
919	63	32P 9102	Nac. Power Plant Inst. L.H. Inbd.
919	64	32P 9103	Nac. Power Plant Inst. R.H. Inbd.
919	65	32P 9104	Nac. Power Plant Inst. R.H. Outbd.
914	66	32P 9111	Turbo Inst. L.H. Outbd.
914	67	32P 9112	Turbo Inst. L.H. Inbd.
914	68	32P 9113	Turbo Inst. R.H. Inbd.
914	69	32P 9114	Turbo Inst. R.H. Outbd.
947	70		Parts List Miscl. Loose Parts
	71		Cancelled
940A	72	32B 9000	Nose Assembly Sta. 0.1 to 4.1
940B	73	32B 9001	Fuse. Assembly Sta. 5.1 to 7.7

Shown here are the Ford part numbers used in April 1944 when Willow Run was at peak production. The item numbers shown on the opposite page were given part numbers and names corresponding to their original uses in the 1941 Consolidated airplane, later expanded and incorporated into the Ford numbering system. The completed items were then assigned to departments inside the Willow Run Bomber Plant. Diagrams on the next two pages show the departments within the factory. After the parts had been manufactured, they were joined into subassemblies for use on the final assembly lines. Much of the equipment on this list was Government Furnished Equipment (GFE) and not manufactured or subcontracted by Ford. GFE included such items as turrets, guns, engines, propellers, and bombsights. Ford installed the GFE items in the final assembly process. (HF.)

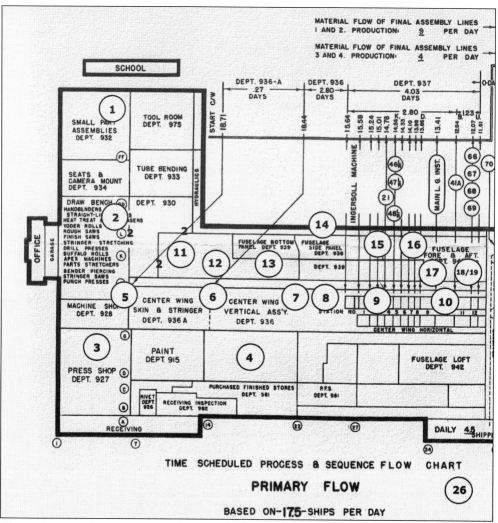

TIME SCHEDULED PROCESS & SEQUENCE FLOW CHART

PRIMARY FLOW

BASED ON—17.5—SHIPS PER DAY

Illustrations on these two pages show the layout of the Willow Run Bomber Plant and the locations of the different manufacturing departments. There are numbers across the bottom of the illustration to identify the frames for east-west location in the factory. Other letters across the left side of this page identify the frames for north-south location. The diagram on this page shows the manufacturing and subassembly areas. On the opposite page are the North Final and South Final Assembly Lines. Between these two areas was the Transfer Bay at frame 40. This allowed component movement between Willow Run's North and South Subassembly Lines. The Transfer Bay also allowed movement to the shipping area for the knockdown kits that would be shipped to B-24 final assembly plants in Tulsa and Fort Worth. (HF.)

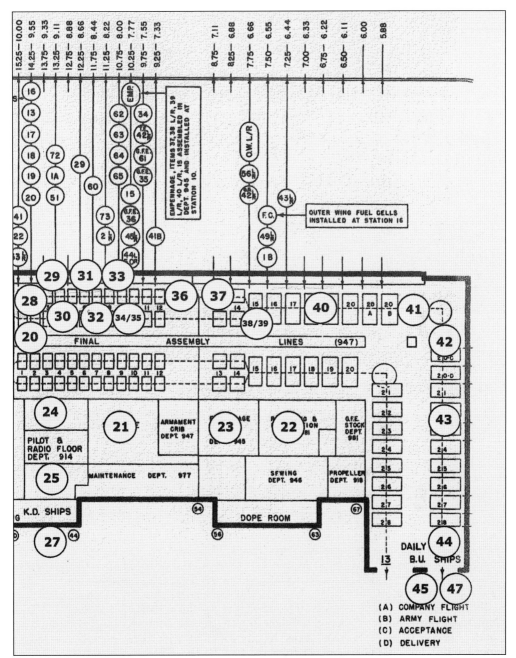

Across the top of both pages are the material flow schedules and the amount of time until that airplane could be flown away and delivered to the Army Air Forces. The circled numbers stacked along the top page correspond to the station on the assembly line where that item was installed. It should be noted that, even after a complete bomber emerged from the factory, it still had five days of inspection and flight-testing before it could be delivered. The miracle of Willow Run can be seen in this schedule. It was only five days from the time the center wing was placed on the assembly line at frame 40 until it came out as a complete bomber. The circled numbers placed throughout the diagrams correspond to the location of the photographs on the following pages. (HF.)

In order to build at a rate of 17.5 bombers per day, the manufacturing area shown here (1) at the west end of the plant produced over 2.5 million parts per day, not including rivets. There were approximately 30,000 unique parts and a total of 150,000 manufactured parts in every Ford-built B-24. Many of the same part numbers were used in assorted locations in each B-24.

The Yoder Roller Machine shown here was located in the Draw Bench Department No. 930 (2). This machine produced wing stringers, stiffeners, and similar parts. This machine, along with Buffalo Rollers and Apex machines, turned out pieces to be used to in the manufacture of 1,700 parts per day. This work had to be scheduled so it would arrive at the assembly stations at the proper time to maintain production rates.

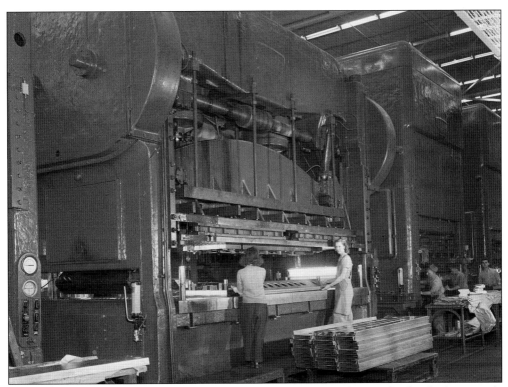

Shown here are women operating a metal stamping press, one of many jobs that had previously been defined as "male only" jobs. The press was located in the Press Shop Department No. 927 (3). The 60 heavy presses in this department produced 4,500 different parts. Ford had mastered the use of hard steel dies in producing aluminum stampings to exacting standards. These allowed output of identical and thus interchangeable parts.

The Press Department (4) could only produce 150 different parts each day. This meant it was necessary to schedule press operations 30 days in advance. If the Army wanted 460 bombers in the next 30 days, it would be necessary to produce 460 parts from one press setup. Shown here are identical parts produced in one day that would be used as a float to maintain production.

This photograph, with a view looking west through the South Subassembly Bay, shows the entire center wing assembly area that included Departments 936, 936A, and 937 (5). The center wing was the most complex assembly manufactured at Willow Run. In the foreground are the prefabrication benches. The benches have layout jigs set for every part to insure identical assemblies were produced. These assemblies were installed inside a center wing as it was being built. Behind the layout benches, the completed inventory was stored in front of the center wing vertical assemblies. There were 43 of these 27-ton vertical fixtures; 43 center wings were under some form of manufacture at all times in order to feed the aircraft production rate of one bomber per hour. Behind the vertical fixtures in the center are the wing machining combine and the final wing dress-up stations.

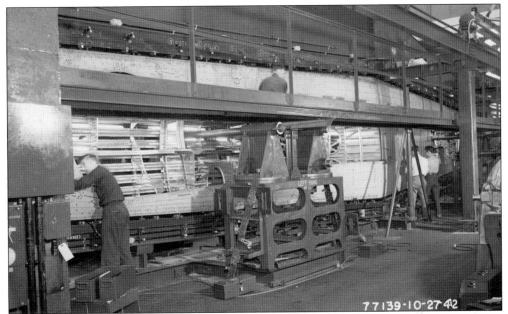

The precision-manufactured spars, stringers, and bulkheads were designed to mate precisely with adjoining parts. They were placed vertically in 27-ton cast-iron and steel center wing assembly fixtures. To save time, the fixtures opened at the top to allow extraction of the completed center wing without disassembly of the fixture. These were manufactured in the section labeled 6 on the drawing.

The metal skin of the center wing was then riveted to the stringers in this fixture. Note the curved uprights in the fixture on which the stringers rest to ensure the proper contour of the wing. At Willow Run, no expense was spared to get the best tools and fixtures to facilitate precision assembly. The riveters in this scene are standing on an elevator platform that raises them to higher levels to complete the riveting job in area 7.

A controversial gigantic Ingersoll machining combine performed 26 almost simultaneous boring and milling operations on the center wing in 45 minutes to tolerances of 2/10,000 (.0002) of an inch. This machine alone cost over $250,000 in 1941 dollars, which the Army considered excessive at the start of bomber manufacturing; however, the combine machine saved hundreds of man-hours in the machining operation and produced identical center wings (8).

After the center wing left the machining combine, it was placed on a horizontal moving assembly line in Department 937 (9). It moved through 12 workstations, and at each station, parts, such as turbochargers and fuel lines, were mounted on the wing. The wing was maintained at standing room height in order to facilitate the installation of components at each station.

At the final station in Department 937 (10), the center wing dress-up process was completed and the wing was elevated to flow into the start of final assembly. When the wing reached this point, it was a nearly complete wing assembly with landing gear, including wheels, tires, brakes, turbochargers, hydraulic lines, and cowl fairing, with all electrical wiring bundles and connectors installed.

Shown here is the North Subassembly line at the start of the fore-and-aft fuselage assemblies (11). Here, parts were organized into a series of buildup stations where parts were added to produce larger components that would be used to complete fuselage sections. The parts would move from the bottom to the top, where they would join the final assembly.

The start of the forward fuselage nose section in Department 938 (12) is shown, with the bombardier and navigator's window visible at left in the picture. Ford again used fixtures to hold stringers and spacers in exact positions while the aluminum skin was attached. The completed side walls would then be moved to dress-up stations where essential equipment and accessories would be installed.

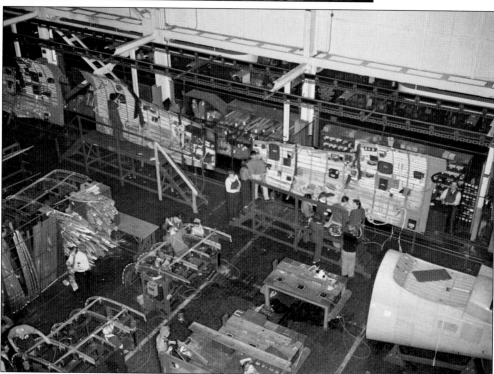

The forward fuselage side walls moved into Department 938 (13), where all of the interior equipment was installed. By having the fuselage section broken into four separate pieces—left and right side walls, plus top and bottom sections—workers had access to rapidly install parts. This avoided the crowded conditions that would be found in a complete forward fuselage, where worker access to interior walls would be limited due to crowded conditions.

In Department 939 (14), rear fuselage sections were being assembled on a parallel line south of the forward nose section line. The fixtures shown here used the same assembling process as the forward fuselage, with side walls open on both sides to allow access by more workers in order to rapidly install needed equipment. In 10 days, this would be part of a flyable B-24.

In Department 939 (15), rear fuselage sections (foreground) on the South Subassembly Line and nose fuselage sections (background) have had the side, floor, and top sections joined to form a fuselage subassembly. Note that bands have been placed around these fuselage sections. In the foreground is a rear fuselage assembly that had been attached to one of the overhead cranes that ran throughout the factory to lift subassemblies out of fixtures.

The completed forward fuselage nose assemblies are shown here in Department 940 (16). The dark metal bands located in front of and behind the pilot's greenhouse were used to lift the completed nose section onto the final assembly line. In the foreground, the number on the forward fuselage just above the radio operator's window aft of the cockpit is the Consolidated Aircraft control number, 1374.

In Department 940 (17), rear fuselage sections neared completion. Here, major components would be attached, such as the tail cone, item 33, which included the horizontal stabilizer twin tail attachment points and tail gunner compartment. The rear fuselages in the upper part of this picture do not have tail cones, and in the foreground, workers put the finishing details on the tail cone installation.

This photograph shows the rear of the fuselage nose section with a variety of capped hydraulic fittings, control cables, and electrical connections in Department 939 (18), which are set up for rapid linkup to the center wing section, where they were mated on the final assembly line. This forward fuselage section is five days from being a flyable airplane.

There was a stock crib for every two workstations (19). The stock crib maintained a 10-day parts float to insure the line was never shut down for lack of parts. Workers had to sign for parts, and if a part was damaged, it had to be turned in before another one would be issued. Parts inventory information was fed into IBM tabulating machines for inventory control. When parts supply fell below a 10-day supply on hand, the parts replacement moved to a higher priority. There were various colors to indicate how short the supply had become and where efforts needed to be focused.

In this photograph is the North South Transfer Bay (at frame 40) where nose and rear fuselage sections were positioned for North Final Assembly Bay operations. At center is a turntable with rollers to feed subassemblies onto the structure that separated the North and South Final Assembly Lines (20) in the North Final Assembly Bay and feed the subassemblies to their attachment stations.

This photograph shows engine nacelle buildup. In Department 919 (21), complete cowlings, ductwork, shrouds, exhaust, and engines were built into these ready-to-install nacelles. However, under Bricker's 505 program, the building of nacelles was transferred to the Lincoln plant in Detroit. Pratt & Whitney–licensed R-1830 engines were shipped directly to the Lincoln plant from Chevrolet's Tonawanda, New York, plant and Buick's Melrose, Illinois, plant. The Lincoln plant was shipping over 100 complete, ready-to-install nacelles per day to Willow Run.

This photograph shows spars and bulkheads for the outer wing being held in place in the vertical fixture (22). Aluminum was then attached to form the outer wing. Again, as part of Bricker's 505 program, the outer wing assembly operations were outsourced to Ford's Highland Park, Michigan, farm tractor plant and then shipped to Willow Run ready for installation.

This photograph shows women workers in Department 912 (23) completing the twin vertical stabilizer and elevator assembly. The rudders, elevator, and horizontal and vertical stabilizers were put together as a single unit to minimize work on the final assembly line. Again, as part of Bricker's 505 program, the tail stabilizer operations were moved to Ford's Rouge Complex and shipped in trucks to Willow Run over what is now known as I-94.

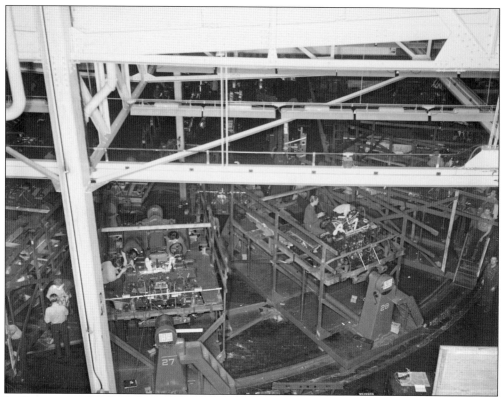

In Department 914 (24), the complex pilots' cockpit was assembled on a moving carousel. The carousel slowly rotated from station to station where parts such as pilot's yokes, rudder pedals, throttles, and engine and flight instrument panels were installed. Once the cockpit was complete, it was lifted by crane and transferred to Department 939 where it was installed in the completed cockpit assembly.

Again in Department 914 (25), the cockpit's center pedestal can be seen in some detail. The women are shown installing equipment and furnishings to the cockpit. Women were working both above and below the cockpit floor, installing and stringing cables that ran between the levers in the cockpit and the engine nacelles located on the wing.

A complete collection of major parts that make up a knockdown B-24 kit can be seen in this photograph taken on the Willow Run ramp (26) in the winter of 1944. Ford shipped 50 of these knockdown kits per month to both Consolidated Aircraft Corporation in Fort Worth, Texas, and Aircraft Corporation at Tulsa, Oklahoma, where Douglas Aircraft and Consolidated Aircraft performed final assembly operations.

This photograph shows the last of 1,893 knockdowns loaded for shipment on July 7, 1944 (section 27). The knockdowns were shipped via specially designed Ford trucks. Trucks were used to avoid possible damage, sabotage, and delays that could have resulted in shipping by railroad. The trucks drove nonstop with two drivers and made the trip in two days, compared to six days on the railroad.

This photograph shows the center wing section being loaded on the North Final Assembly Line at Station One, the first of 28 stations in Department 947 (28). Women workers can be seen placing item 22, life-raft storage bay, on top of the wing. At this point of assembly, there were four parallel lines, and every four hours, the line would move ahead one station. This wing is five days from being part of a completed airplane.

The photograph shows the work that had been done under the wing installing bomb bay side walls. In the background, complete nose and tail fuselage subassemblies can be seen on the transfer bay at frame 40 (29). Rollers riding in rails at the wing tips supported the center wing. The rollers were attached to a chain drive that moved the wing every four hours.

Shown here in the upper right at Station 4 (30) is the nose section, item 72, in the installation cradle. The cradles were an example of Ford's labor-saving devices that sped up the final assembly process. The nose section would slide under the cradle where workers would attach cables to eyelets on the metal straps that wrapped around the nose section and move it into position.

In area 31, with the nose section in the upper right now attached to the cradle, the cradle moved either left or right to join the center wing assembly on parallel final assembly lines. At Station 4, when the nose was attached to the center wing, the airplane was only four days from being a complete flyable airplane.

The center wing came out of Station 4 where the nose section and nose turret mount were attached to the center wing. This assembly moved into Station 5 where the hydraulic reservoir was installed and Station 6 (32) where the nose landing gear was installed. In the background is a cradle similar to the nose cradle, ready to receive the rear fuselage, item 73, which will be attached at Station 7.

This photograph shows the floor level at Station 7 (33) where the rear fuselage subassembly was being swung into position for attachment to the rear of the center wing. The men shown here will assist in positioning the rear fuselage onto the fixture on the floor and then push it forward for mating to the center wing.

Here, the partially assembled airplane moved into engine nacelle installation at Station 9 (34). The airplane now rides on its own landing gear. The four engines manufactured by General Motors, built into finished nacelles at the Lincoln plant, are about to be installed on a B-24 on a Ford assembly line. Such was the coordinated auto industry involvement in airplane manufacturing in World War II.

The completed engine nacelle was dropped into position for installation on the center wing in area 35. The men are guiding the nacelle to its attachment point. At peak production, workers were installing an engine in 15 minutes. There was still much more work to be done to complete the installation, such as attaching all cables and plumbing to the engines. At this point the airplane was three days from being a completed Ford B-24.

At Station 11 (36), on the second airplane from the front, items 38, 39, 40, and 41 are being installed as a completed empennage. The woman working in the overhead crane is lowering the tail assembly precisely into place. Due to the precision fit of parts shipped from the Ford Rouge Complex, the installation process took less than half an hour to complete.

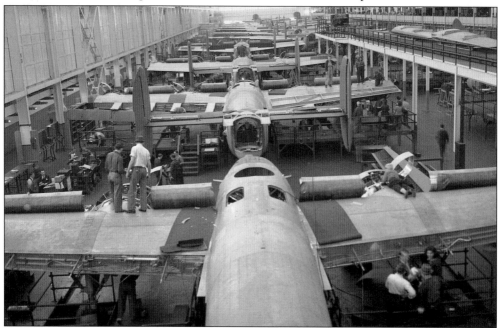

The airplane can be seen as it moved through the last of the four-hour stations—11, 12, 13, and 14, where parts, such as the center wing flaps manufactured by the Gibson Refrigerator Company and the trailing edge fairings built by E.G. Budd Company, were installed (37). Outer wings that will be installed at Station 15 can be seen on the elevated structure at right.

In the center of this photograph, in area 38, World War I ace Eddie Rickenbacker is shown with some of the little people. Henry Ford employed little people because they could get into the cramped spaces inside the center wing. They attached the outer wings and were proud of their role in building the bombers at Willow Run. For the first time in their lives, they were being recognized for what they could do as opposed to how they looked.

This image shows the outer wing being attached to the center wing at Station 15 (39). Peering out of the center wing is one of the little people waiting to fasten the outer wing. At Station 15, the airplanes moved to the center of the assembly bay and transitioned through two-hour stations. This B-24 is less than a day and a half from being a complete airplane.

This photograph shows nearly completed B-24s moving through Stations 16 to 20 in area 40. These are each two-hour stations. Time spent at these stations was used to finish installation of items attached at earlier stations, such as final hookup of all wiring and plumbing. At Station 20, the Hamilton Standard propellers manufactured by Kelvinator Refrigerator Company were installed, and the airplane was 16 hours from being complete.

The airplane was pulled onto the famous Ford Willow Run turntable by a Ford tractor that has been modified into an airplane tug. Originally, as the plant was designed, the airplane would have been pulled straight out of the factory at this time. However, additional demands placed by the Army in October 1941 caused Ford to lengthen the assembly line with a 90-degree turn to the south (41).

The B-24 moved off the 90-degree turntable and the center island structure designed to support the roof when the plant was modified in October 1941. Seen at left is the South, East-West, Final Assembly Line and on the right is the North, East-West, Final Assembly Line. At any one time, 90 B-24s were on the final assembly line (42).

The airplane is shown here in the paint booth at Station 21 (43). The paint booths were designed to allow a rapid application of the Army's olive drab camouflage paint scheme. Built-in elevator platforms allowed the painters to easily move up and down to apply the paint. In December 1943, the Army did away with painting the airplanes to reduce weight and time in assembly. The weight reduction allowed heavier payload, greater speed, and longer range for the completed airplane.

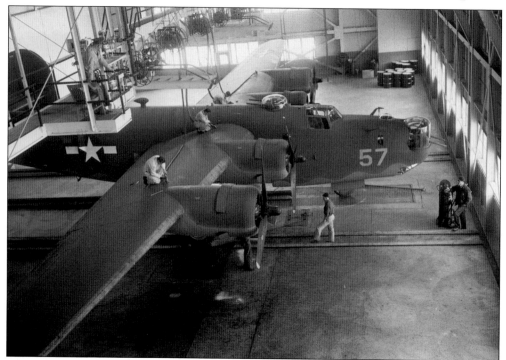

The final operation in the Willow Run factory was adding oil and fuel to the airplane, just as on automotive assembly lines. As the airplane was moved into Station 28 (44), a fireproof barrier was dropped behind it. Then, men would step down from the platform above with hoses and add 100-octane aviation gasoline and oil to the plane. The airplane could then be delivered to the flight test crew for engine run-up and other testing procedures.

In this publicity photograph taken in September 1943, two B-24 bombers emerge from the south-facing doors (45) of the Willow Run Bomber Plant. These airplanes were still five days away from being delivered to the Army at Romulus Army Air Field. Ford employee pilots would fly the airplanes over to Romulus, now called Detroit Metro Airport, to be delivered to the Army's Ferry Command.

The Ford flight-test crew is pictured walking out to a finished B-24 in area 46. Ford was required to fly the airplane for a minimum of 2.5 hours. During these flight tests, engines were shut down and restarted, propellers feathered and unfeathered, and practice bombs dropped to ensure everything functioned normally. These flights had to be done during the daytime and only on good-weather days, which often delayed deliveries. Ford delivered a record 128 bombers to the Army in one day due to backup caused by bad weather.

Shown here in March 1945, in the center of the picture, are Henry Ford II (fifth from left) and Mead Bricker (sixth from left) with a B-24M in area 47. It was the 8,000th B-24 built by Ford at Willow Run. The B-24M was the last large-scale production variant of the Liberator. In March, the Army announced that B-24 assembly would be terminated by August 1945. The end of the war in Europe brought an even more rapid end to production; the contract with Ford was officially terminated on May 31, 1945, and the last airplane delivered on June 28, 1945.

BIBLIOGRAPHY

Beasley, Norman. *Knudsen: A Biography.* New York: McGraw-Hill, 1947.

Brinkley, Douglas. *Wheels for the World: Henry Ford, His Company, and a Century of Progress.* New York: Penguin Group, 2003.

Carew, Michael G. *Becoming the Arsenal: The American Industrial Mobilization for World War II, 1938–1942.* New York: University Press of America, 2010.

Davis, Michael, W.R. *Detroit's Wartime Industry: Arsenal of Democracy.* Charleston, SC: Arcadia Publishing, 2007.

Doolittle, Gen. James H. *I Could Never Be So Lucky Again.* New York: Bantam Books, 1991.

Hamilton, Nigel. *The Mantle of Command: FDR at War, 1941–1942.* New York: Houghton Mifflin Harcourt, 2014.

Holley, Irving E. Jr. *Buying Aircraft: Material Procurement for the Army Air Force.* Washington, DC: Center of Military History United States Army, 1964.

Hyde, Charles K. *Arsenal of Democracy: The America Automobile Industry in World War II.* Detroit: Wayne State University Press, 2013.

Kidder, Warren B. *Willow Run: Colossus of American Industry.* Lansing, MI: self-published, 1994.

Lindbergh, Charles A. *The Wartime Journals of Charles A. Lindbergh.* New York: Harcourt Brace Jovanovich, 1970.

Miller, Donald L. *Masters of the Air: America's Bomber Boys Who Fought the Air War Against Nazi Germany.* New York: Simon & Schuster, 2006.

O'Callaghan, Timothy J. *Ford in the Service of America.* Jefferson, NC: McFarland & Company Inc., 2009.

Peterson, Sarah J. *Planning the Home Front: Building Bombers and Communities at Willow Run.* Chicago: University of Chicago Press, 2013.

Sitterdon, Carlyle. *Aircraft Production Policies Under the National Defense Advisory Commission and Office of Production Management.* Washington, DC: Civilian Production Administration, 1946.

Sorensen, Charles E. *My Forty Years at Ford.* New York: W.W. Norton and Company, 1956.

ABOUT THE YANKEE AIR MUSEUM

Founded in 1981 at Willow Run Airport, the Yankee Air Museum, a nonprofit 501(c)(3) tax-exempt organization, attracts people of all ages. At its headquarters on the east side of Willow Run Airport, it houses a collection of aviation-themed exhibits with many aircraft that shaped history. The museum features a local focus on the Willow Run B-24 Bomber Plant. On October 9, 2004, the World War II hangar that was home to the Yankee Air Museum suffered a fire that destroyed the museum and virtually all of the documents and artifacts. In 2009, to reestablish the museum, the Yankee Air Museum purchased a former aviation technical school building at Willow Run. In 2010, exactly six years after the catastrophic fire, the Yankee Air Museum Collections & Exhibits Building was dedicated and opened to the public. It quickly grew and is now brimming with artifacts. In 2004, when the fire destroyed the museum's home, the Michigan Aerospace Foundation was founded to support capital projects fundraising for the Yankee Air Museum. In 2011, the opportunity arose to save a portion of the historic Willow Run Bomber Plant. Phase 1 fundraising was successful; and in October 2014, purchase of the south end of the Bomber Plant was accomplished. The Yankee Air Museum has a new and historical home rising and when it moves in, it will be the National Museum of Aviation and Technology at Historic Willow Run. Fundraising and restoration of the building continue today with a 2018 goal for completion as the new home of the Yankee Air Museum.

All proceeds from on-site retail sales of this book go towards the future of the Yankee Air Museum. For more information on the Yankee Air Museum, you can visit our website at www.yankeeairmuseum.org. To make a donation that will help restore the bomber plant, go to the Michigan Aerospace Foundation website at savethebomberplant.org.

DISCOVER THOUSANDS OF LOCAL HISTORY BOOKS FEATURING MILLIONS OF VINTAGE IMAGES

Arcadia Publishing, the leading local history publisher in the United States, is committed to making history accessible and meaningful through publishing books that celebrate and preserve the heritage of America's people and places.

Find more books like this at
www.arcadiapublishing.com

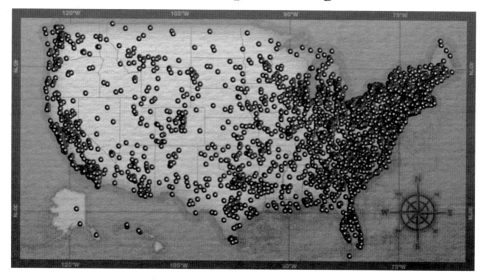

Search for your hometown history, your old stomping grounds, and even your favorite sports team.